# Design and the Social Sciences

# Design and the Social Sciences

## Making connections

**Edited by Jorge Frascara**

London and New York

contemporary trends institute

First published 2002
by Taylor & Francis
11 New Fetter Lane, London EC4P 4EE

Simultaneously published in the USA and Canada
by Taylor & Francis Inc
29 West 35th Street, New York, NY 10001

*Taylor & Francis is an imprint of the Taylor & Francis Group*



Typeset in Sabon by Exe Valley Dataset Ltd, Exeter
Printed and bound in Great Britain by
St Edmundsbury Press, Bury St Edmunds, Suffolk



*British Library Cataloguing in Publication Data*
A catalogue record for this book is available
from the British Library

*Library of Congress Cataloguing in Publication Data*
A catalog record for this book has been requested

ISBN 0–415–27376–5

# Contents

# Notes on contributors

**Rae Bridgman** is an urban anthropologist and visual artist, an Assistant Professor in the Department of City Planning, Faculty of Architecture, and Adjunct Professor in the Department of Anthropology at the University of Manitoba. She is also a Research Associate with the Department of Social Anthropology at York University, Canada. Rae Bridgman has written many articles on homelessness, and is co-author of *Braving the Street: The Anthropology of Homelessness* (New York: Berghahn Books, 1999), and co-editor of *Feminist Fields: Ethnographic Insights* (Peterborough, Ontario: Broadview Press, 1999). She has been documenting the life histories of StreetCity, Strachan House and Savard's since 1994, and as well, has most recently begun to document Eva's Place, transitional housing for street youth in Toronto.

**Ausra Burns** received her Master of Design degree from the University of Alberta in 1997 and is now working on her PhD dissertation, part of an interdisciplinary program. Ausra is currently involved in a community-based design project in New Brunswick, Canada.

In her research and professional work, Ausra aims to correlate knowledge from design and the social sciences, and to integrate theory with practice. She has particular interests in issues of designing for urban public spaces and interdisciplinary collaborative work processes for design projects.

**Peter Dixon** is a Professor in the Department of Psychology at the University of Alberta where he has been since 1981. He obtained his PhD at Carnegie Mellon University in 1979. He publishes in the areas of language, attention, visual perception, motor control, and statistical methods. His recent publications include "Meaning resolution processes for words: a parallel independent model" with Leslie C. Twilley, to be published in *Psychonomic Bulletin & Review*, "Attentional components of partial report", with Robert Gordon, Alice Leung, and Vincent Di Lollo, in *Journal of Experimental Psychology: Human Perception and Performance*, and "Scientific versus statistical inference", with Tenaha O'Reilly, in *Canadian Journal of Experimental Psychology*.

**Maurits Ertsen** was born 1968 in the Netherlands, graduated from Wageningen Agricultural University in 1993, with a major in irrigation, a major in philosophy, and a minor in hydraulic engineering. From March 1994 until June 1995 he worked as research coordinator for the Research Group on Underdevelopment (TEGON) in Wageningen. From September 1995 until November 1996 he was researcher at the Department of Irrigation of WAU on the historical changes in the irrigation system Proyecto Río Dulce (PRD), in northwest Argentina. The first three months of 1997 he took a post-doctoral course in Environmental Public Management. From April 1997 until September 1999 he worked at the Faculty of Technology, Policy and Management of Delft University of Technology, Holland, conducting research on possibilities to integrate education in the social sciences into the educational programs of the engineering faculties.

Since January 1999 he has also been working at the Department of Land and Water Management at the faculty of Civil Engineering, conducting a PhD research project on the historical development of irrigation technology in the Netherlands' East Indies. This project is conducted in cooperation with the Department of History of Technology of the Faculty of Technology, Policy and Management. From September 1999 onwards, he is combining this research with the coordinatorship of project education at the faculty of Civil Engineering of DUT.

**Jorge Frascara** is Professor of Visual Communication Design at the University of Alberta. He is a Fellow of the Society of Graphic Designers of Canada and an advisor for five professional journals and for the Design Department at Carnegie Mellon University. He is a member of the ISO Technical Committee for Graphic Symbols and an International Fellow of the Society for the Science of Design of Japan.

Among other positions, he has been Chairman of Art and Design at the University of Alberta and President of Icograda, the International Council of Graphic Design Associations. He has organized several international conferences in North America, Europe, Africa and Latin America and has published extensively on design, communication theory and design education. His professional experience includes illustration, film animation, advertising and graphic design, and now concentrates on research and development of visual communications for safety and other social concerns.

**Patrick W. Jordan** is a Vice-President of Symbian, where he is also head of design. Symbian is a joint venture mobile communications company jointly owned by Motorola, Nokia, Psion, Ericsson and Panasonic. Symbian also licenses to Kenwood, Philips, Sanyo and Sony. Pat is also President of the Contemporary Trends Institute [CTI], an international trends and branding consultancy. Formerly, he was Head of Human

Factors at Philips Electronics, where he later became Director of the Trends and Identity Unit.

Pat has been invited to lecture at conferences and seminars all over the world. He has over seventy publications in peer reviewed journals, books and conference proceedings, has written or edited five books and is currently Europe's best-selling author in his field. His books include *Designing Pleasurable Products* (Taylor & Francis 2000), the standard text for affective design, and *An Introduction to Usability* (Taylor & Francis 1998), which has spent over two years in the Amazon.com design bestsellers top 10, reaching the no. 1 position earlier this year. His forthcoming book *Supertrends* will be published by Kogan Page in 2002.

Dr. Jordan has a visiting lectureship at the London College of Fashion and is a guest lecturer at University College London and Carnegie-Mellon, Brunel and Loughborough Universities. He has won numerous professional awards for design and related activities.

**Carrie A. Lavis** completed her PhD in Psychology at the University of Alberta. Her research interests include the effects of mood states on behavioral (i.e. workplace performance and productivity) and attitudinal (i.e. responses to persuasion) outcomes, and the impact of learning technology initiatives on professional development. She has published in *Nature*, and will be working with Julian Barling on a project funded by the Center for Automotive Materials and Manufacturing (CAMM) entitled "Transformational leadership, high performance work systems, employee performance and productivity in the automotive sector." She will be teaching a course in Organizational Behavior during the winter term at Queen's University, Kington, Ontario.

**Bernd Meurer** is Professor Emeritus in Design at the Fachhochschule Darmstadt and Director of the Laboratorium der Zivilisation, Darmstadt. He has taught at universities in Germany, the USA and Japan. He is a member of the Deutscher Werkbund, of the City Development Council of the City of Darmstadt, of the Scientific Commitee of the Istituto Europeo di Design at Cagliari, Italy, of the foundation committee for the Bauhaus Kollege Dessau, of the European Committee "The New Academy" at the European Leage of Institutes of the Arts (ELIA), and founding member of The Society for the Science of Design Studies in Tokyo. He has published several books and articles, most notably for English readers, *The Future of Space*, a book produced as a result of one of a series of conferences Meurer organized around interdisciplinary cooperation, technology and sustainable development.

**Sean E. Moore** is a PhD candidate in the Department of Psychology at the University of Alberta. He received his BA from Wilfrid Laurier University in 1997. His research interests span a range of topics in social psychology including the study of affective states and their influence on cognitive processes, attitudes and attitude change, and the application of social psychology to organizational behavior and marketing research.

**Patricia Neafsey** is Associate Professor at the University of Connecticut's School of Nursing. Dr. Neafsey holds an MS in nutritional biochemistry from Cornell University, a PhD in Pharmaceutical Science from the University of Connecticut, and has completed post-doctoral work at Tufts University. She has authored more than 32 papers on nutrition, pharmacology, and the use of interactive computer technology in nursing education and patient dose risk reduction. Her recent work involves a research collaboration to measure the effectiveness of an interactive, animated educational software program for reducing high-risk self-medication behaviors in a population of older adults.

**Christena Nippert-Eng** is Assistant Professor of Sociology at the Illinois Institute of Technology in Chicago. Her research focuses on issues of home, work, time, space, cognition, technology and privacy, and has been the focus of numerous national and international radio and television shows, magazine and newspaper stories. Dr. Nippert-Eng teaches lecture-, film- and project-based courses at IIT, and has received numerous teaching and service awards. She enjoys collaborations with faculty at the Institute for Law and the Workplace at the Chicago-Kent School of Law, IIT's Institute of Design, and the Graduate Program in Public Administration. In addition to her work with non-profit organizations Dr. Nippert-Eng designs and conducts field research for industry. Dr. Nippert-Eng is a Founding member of the Board of Directors for the Center for Ethnographic and Documentary Research (CeDoc) a coalition of academic and industrial partners ranging from social scientists and designers to film-makers and photographers.

**Tenaha O'Reilly** is a graduate student in the Department of Psychology at the University of Alberta where he has been since 1997. He is interested in the problem of learning and using operating procedures for complex procedures. His recent publications include "Procedures are only skin deep: the effects of surface content and surface appearance on the transfer of prior knowledge in complex device operation," in the *Proceedings of the 1999 Cognitive Science Meeting*.

**Romedi Passini** holds a degree in Architecture from the Eidgenössische Technische Hochschule, Zürich, Switzerland, and a PhD in Environmental Psychology from Pennsylvania State University. He has worked professionally as an architect and has been a consultant on wayfinding, signage and mobility for various governmental and private agencies.

His academic experience has concentrated on the University of Montreal, where he taught courses in environmental psychology, spatial orientation and wayfinding, research methodology and design studio. His research experience has been in the field of environmental psychology and architectural design, with specialization in wayfinding, spatial orientation, environmental perception and cognition and including studies concerning mobility and spatial orientation of the visually handicapped and the elderly population. His publications are too numerous to do justice to them here but some of his most recent titles include: "Visual Information for Everyday Use: Design and Research Perspectives," "Wayfinding and Information Processing in Dementia," "Information Design: An Old Hag in Fashionable Clothes?," "Wayfinding Design: Logic, Application and Some Thoughts on Universality," "Wayfinding in Dementia: Some Research Finding and a New Look at Design." He has published in very diverse journals, including the *Journal of Architecture and Planning Research*, the *Journal of Clinical and Experimental Neuro Psychology* and the *Journal of the International Neuropsychological Society*.

**Sharon Helmer Poggenpohl** is Associate Professor at the Institute of Design, Illinois Institute of Technology, where she coordinates the Doctoral Program in Design. She is editor and publisher of the interdisciplinary journal *Visible Language*, from where she has had a sustained influence on the development of design theory. She has published many articles in leading journals. A recent book *Design Innovation for Global Competition* was published in 1995 in English and Chinese in collaboration with colleagues at the Institute of Design, IIT.

**Desmond Rochfort** is President of the Alberta College of Art and Design in Calgary. Born in Zimbawe and educated in England as a painter, he obtained his PhD for his research on the politics and art in Mexican mural painter Siqueiros. His second book on Mexican painting has been published in England, the USA and Mexico. He was Chairman of the Department of Art and Design at the University of Alberta until June 1999, and was at the center of the introduction of the Bachelor of Design degree and of the crafting of the pathways system for that program. The Design and the Social Sciences conference was conceived to contribute to the new pathway that Art and Design opened with the departments of Anthropology, Psychology, and Sociology.

**Elizabeth B.-N. Sanders** is the President of SonicRim. With undergraduate degrees in psychology and anthropology, and a PhD in Cognitive Psychology, Liz brings a human-centered perspective to everything she touches. Focusing on the user, and collaborating with others who understand the market, she explored and perfected innovative design research techniques at Fitch for seventeen years. There she headed up the Research and Planning office, helping to establish it as the leader in the field. Now, at SonicRim, Liz is expanding the scope of applying her expertise and intuition to domains beyond design research.

Liz's numerous design awards, publications, patents, conference presentations, and her proven track record in the marketplace have established her as a leader in the field of design research. Her client relationships have included 3M, AT&T, Apple, Baxter, Coca Cola, Compaq, Hasbro, IBM, Iomega, Microsoft, Motorola, Procter & Gamble, Siemens, Steelcase, Texas Instruments, Thermos, Toro, and Xerox.

Through the Industrial Design Department at Ohio State University (OSU), where she has been constructing learning environments for interdisciplinary graduate student teams for ten years, Dr. Sanders has gained invaluable experience in facilitating hands-on learning. She is an adjunct faculty member in both the Industrial Design Department and in the Industrial and Systems Engineering Department at OSU.

**Rob Shields** holds an MA in Sociology and a DPhil in Urban and Regional Studies. He is Professor of Sociology and Anthropology at Carleton University, Ottawa and was a Commonwealth Scholar at Sussex University, and also lectured at Lancaster University, in England. Originally trained as an architect, his publications include *Lefebvre: Love and Struggle* (Routledge); *Places on the Margin: Alternative Geographies of Modernity* (Routledge 1991; Outstanding Publication of the Year Award); and edited best-selling collections such as *Lifestyle Shopping* (Routledge 1992), *Cultures of Internet* (Sage 1996) and *Social Engineering* (Carleton 1996). Consultancies have included the ESRC "Virtual Society?" Program Proposal; work in the UK for DoE on motivating innovation in construction and evaluation guidelines for the DoE PiT program and in Canada for the NRC's Institute for Research on Construction. Current research includes internet communities; virtuality; an ethnography of innovation and change on construction sites (for SSHRC Canada). He heads a major research program on the Public Service and the Knowledge Based Economy (SSHRC Canada Strategic Theme) and is the founding editor of *Space and Culture*.

**Robert C. Sinclair** is President of Sinclair & Associates, Organizational Consultants, and was Associate Professor in the Department of Psychology at the University of Alberta. His research interests center on the relations between cognitive and affective processes, and include motivation, cognitive approaches to person perception, social justice, methodology, program evaluation, biases in survey/questionnaire responses, and applications in the areas of organizational behavior and marketing. He has taught at Central Michigan University and at Pennsylvania State University. His recent publications include "An electoral butterfly effect" (*Nature* 2000), with M. M. Mark, S. E. Moore, C. A. Lavis and A. S. Soldat, "Colors, smiles, and frowns: external affective cues can directly affect responses to persuasive communications in a mood-like manner without affecting mood" (*Social Cognition* 2001), with A. S. Soldat, and "Globalisation, diet, and health in the micro-states of the South Pacific: an example from Tonga" (*International Bulletin of the*

*World Health Organisation*, in press), with M. Evans, C. Fusimalohi and V. Laiva'a.

**Alexander S. Soldat** has received a PhD in Psychology from the University of Alberta and is now Assistant Professor at Idaho State University. His major research interests are: models of the effects of non-mood-related affective cues on social/cognitive processes; attitudes and persuasion; methodology; and applications of social cognition.

**Louise St Pierre** is Assistant Professor of Industrial Design at the University of Washington. She has taught industrial design for several years at the Emily Carr College of Art and Design in Vancouver, Canada. She has practiced both industrial and exhibition design and her work has received several Canadian design awards. Her research work follows her interest in objects and in their social relevance. Through her relationships with interdisciplinary organizations such as CEEDS (the University of Washington's Center for Environment, Education, and Design) she collaborates with designers, social scientists, educators, and environmental psychologists in defining physical objects which can support educational objectives.

**Zoe Strickler**, MDes in Visual Communication Design (University of Alberta), is Design Research Specialist at the University of Connecticut's School of Nursing. During the 1980s she worked as a graphic designer for the firms Eaton & Associates Design and Cohen Little Design in Minneapolis where her work won national awards for excellence. Her research interests link theories of behavioral change to health and safety communication applications for specific audiences. She is currently working as a design researcher in the School of Nursing at the University of Connecticut. She has held academic appointments at the Minneapolis College of Art and Design, College of St. Catherine, and the University of Connecticut.

**Rosalind A. Sydie** is Professor of Sociology at the University of Alberta. She is primarily a theorist with interests in art, culture and gender. Her interest in the phenomenology of space and the built environment is long standing, and in part arises from early associations with some architects during her time at the University of Liverpool.

**Jim Wilson*** at the time of writing this article was the Manager of Human Factors & Ergonomics in Motorola's Communications Enterprise, Fort Lauderdale, Florida. He had 19 years of basic and applied experience in Cognitive and Human Factors Psychology, most of it applied to "smart" consumer and commercial product design, development, and evaluation. He earned the MS in Developmental Psychology and PhD in Cognitive Psychology/Human Factors at the Pennsylvania State University, and a

BA in Psychology at the University of Maryland. He taught undergraduate and graduate courses in Human Factors, Product Design, Psychology, and Statistics at the University of Rochester, SUNY Buffalo, and the Pennsylvania State University, and also worked as a graduate thesis advisor on projects in Psychology, Human Factors, Computer Science, Electrical Engineering, and Industrial Design. He published in the areas of user interface design, rapid prototyping, the product development process, the software development process, and in selective visual attention. He was with Motorola for six years, prior to that he worked for over eight years with the Eastman Kodak Company in Rochester, New York.

**Dietmar R. Winkler** was born in Germany, and since 1960 has been combining professional design practice with teaching of design and communication subjects. In 1998 he became Director of the School of Art and Design at the University of Illinois at Urbana-Champaign, leaving his Hall Chair at the Kansas City Art Institute, where he directed the Center for Form, Image and Text. For many years, he was a member of the Design Department and an adjunct in the cognitive science program of the Psychology Department at the University of Massachusetts Dartmouth. His projects have focused on visualization of complex and incongruent data, and multi-entry hypertext environments. He was also responsible for the initiation of a multi-disciplinary project involving the departments of education, science, communication, journalism, design, art, political science, history, sociology, psychology, and cognitive science. He has been associated to several post secondary institutions and is a member of the editorial board of the *Visible Language* journal.

In design practice, he has worked for Brandeis University, Harvard Business School, Massachusetts Institute of Technology, University of Massachusetts Dartmouth, as well as the WGBH Educational Foundation. His design work has been awarded, exhibited, and published by the Art Directors Clubs of Boston, New York, and St. Louis, the Type Directors Club of New York, and the American Institute of Graphic Arts.

---

*Jim Wilson passed away between the conference and the publication of this book. Although I only met Jim at the conference and through our correspondence leading to it, I liked him and admired him very much. Forthcoming, to the point, efficient, friendly and generous, we owe him a good part of the quality of the conference and of this publication. It was him who actively put me in touch with some of the best contributors as soon as I invited him. Thankful, I pay homage to a remarkable person.

J.F.

# Preface

The main objective of this project is to identify and define possible directions for interdisciplinary work connecting design and the social sciences regarding the conception, production and use of objects, environments and communications. The conference held in Edmonton was intended to critically discuss both existing realities and the benefits that a more interdisciplinary action could bring to research, practice and education, challenging, at the same time, existing notions of disciplinary competence.

The design discipline has developed in recent years from an exclusive concentration on the design of objects, environments and communications toward an expansion of its field to include the design of processes, services, structures and systems, and to the creation and promotion of ideas and principles; in sum, to a series of activities that could be defined as the design of the contexts within which traditional design operates. These contexts involve the critical consideration of social, cultural, economic, technical and environmental concerns, and map out a broad terrain for designing and manufacturing. The need for people prepared to work at this level is growing every day, as hitherto hidden dimensions of the economy—such as the cost of health care and illiteracy—and the challenges brought by international markets, pose heavy demands on creativity and efficiency to any human group.

Product development is now far from being the province of the individual craftsman, manager or manufacturer: the conception, production, distribution and use of products have become complex parts of corporate strategies, and include extensive research based on marketing, but also on anthropology, psychology and sociology. The same could be said of instructional and educational materials: this is not any longer the terrain of teachers specialized in specific content areas. The development of teaching aids calls for knowledge in cognitive, developmental and perceptual psychology, as well as design-specific knowledge of media, production and components; the evaluation of the effectiveness of any design product or idea makes use of knowledge developed within a variety of fields. Manufacturers such as Philips, and institutions such as the International Standards Organization (ISO) have for over twenty years gathered together designers and social scientists around the challenges posed by the design of products and communications. Design education, however, has remained behind, still today following art- and craft-based models, at best with some unconnected components provided by other disciplines.

Responding to this, the University of Alberta has recently launched a Bachelor of Design program that formally integrates courses in anthropology, psychology and sociology, aiming at the formation of a designer with extensive awareness and systematic formation in the methods of those disciplines, so as to reach a better understanding of people—the users of design—in addition to the traditional components of design education. The Bachelor of Design with Pathways, which already had pathways in Business, Computing Science and Engineering, initiated a pathway through the social sciences in September 1999.

The conference held in 1999 included presentations and four working groups. Each one of the working groups had a specific profile: design and the social sciences in the university, design and the social sciences in industry, design and the social sciences in research and interdisciplinary cooperation. While the three social sciences have much to offer to designers from their knowledge, methods and studies of different aspects of humanity, design in turn can offer the social sciences—which normally deal with the analysis of existing objects and situations—an opportunity for intense interaction with the conception and production of the artificial world.

Two sessions of working groups were programmed. Their aim was to open lines of communication and reflection, to identify relevant issues for discussion and to begin to deal with them.

The objectives of the working groups were: (a) to generate recommendations for ways to develop and improve communication and cooperation between design and the social sciences in industry and the university; (b) to identify areas of action; (c) to describe past and current experiences and systems of cooperation; and (d) to outline future agendas.

The proposed framework for the discussions of the working sessions were defined by five oppositions:

1 *Conversation vs. debate.* It was proposed to use the conversation as a model, not the debate. In conversation, people converge on a subject without fixed positions, without polarizations. It is different in a debate, where normally two positions are taken, where differences within groups are obliterated, and where the objective is to win. In the conversation, the objective is to exchange perceptions in order to enjoy communication and improve understanding.
2 *Understanding vs. agreeing.* Another aim of the working groups was to center on understanding, rather than on agreeing. Participants were asked to ask questions directed at clarifying and elaborating, so as to ensure understanding of every issue raised, without being pressed to arrive at a consensus position as a group.

3 *Multiplicity and complexity vs. unity and simplicity.* The meeting was geared toward identifying a multiplicity of options rather than toward the creation of a unified conception as to how the relation between design and the social sciences should be.
4 *Reflection vs. censorship.* The main aim of the meeting was to take advantage of the personal presence of so many people, and to avoid dismissing any points raised. It is necessary to allow time for ideas to become well articulated; therefore working groups were asked to allow imperfect ideas to be expressed. This was a fair price when trying to identify issues that have not been articulated before.
5 In sum, the working groups were asked to identify possible, desirable and necessary directions regarding the working relation between design and the social sciences, in both industry and the university.

The conference closed with the reports of the working sessions, opening the dialogue between designers and social scientists toward mapping the terrain for future cooperation, proposing strategies for approaches to the interaction between design and the social sciences in research, education and practice. This book documents the event and hopefully opens further possibilities for future dialogue.

The material selected here is organized in a continuous sequence. The spirit of the conference and this publication is one of exchange and integration. Although speakers came from different fields, the design focus created a center of convergence where different ideas merge. Without creating boundaries, sections or separations, the articles have been arranged beginning with those by professionals who work in industry (Sanders, Jordan, and Wilson). These are followed by speakers who represent education, research and practice in design, from the more abstract and general articles to those that report on case studies (Frascara, Meurer, Winkler, Poggenpohl, Burns, Passini, Bridgman, Strickler and Neafsey, StPierre, and Ertsen). Then an educational administrator (Rochfort); followed by social scientists (Dixon/O'Reilly, and Sinclair/ Moore/Lavis/Soldat, Psychology; Sydie, Shields, and Nippert-Eng, Sociology). The book closes with the reports of the four working sessions, and a brief conclusion.

JORGE FRASCARA
Editor and Conference Chair
University of Alberta
Edmonton, Canada, August 2001

# Acknowledgements

My first thanks go to the University of Alberta, whose officers at several levels supported the development of the conference that made this book possible. The conference organizing committee I chaired included Desmond Rochfort (Chairman of Art and Design at the time of planning the conference), Eric Higgs (Anthropology), Rosalind Sydie (Sociology) and Connie Varnhagen (Psychology). Susan Colberg (Art and Design) and Bonnie Sadler Takach (Art and Design) organized an international project and an exhibition, thus extending the reach of the effort. Raúl Martínez (Art and Design), a graduate student at the time, deployed endless amounts of energy and resourcefulness all along the organization process down to the last detail and at many levels. I thank all for the time, knowledge and skills they put into shaping the concept and making it possible. My thanks also to Mariana Waisman, for her diligent and efficient assistance in the preparation of the final manuscript for publication.

The speakers created a stimulating environment, and engaged in interaction with the participants with a seldom seen intensity. I thank them all, particularly those who so efficiently responded to the organizational needs of the conference and the logistical and editorial details of producing the book, making my tasks more possible. My special thanks to Pat Jordan, who from the very beginning contributed so generously, and who is chiefly responsible for the possibility to bring this book out to the public.

J.F.

# 1 From user-centered to participatory design approaches

*Elizabeth B.-N. Sanders*

## Background

The integration of design with the applied social sciences is relatively new. Design firms began experimenting with the social sciences in the early 1980s. The experiment was design-driven, with social scientists being brought in to serve the design process.

As a social scientist trained both in psychology and anthropology, I was one of these "experiments." I began to serve the design process in 1982. In the 1980s I played the role of the human factors practitioner, or "user advocate." My role was to know the user and to translate that knowing into principles and prescriptions that the designers with whom I worked could understand and use. We called this the user-centered design process. As I learned ways to help make products and information systems more usable, I also studied the designers, especially the ways they visually communicated with each other.

## User-centered design process

In the user-centered design process, we are focused on the thing being designed (e.g. the object, communication, space, interface, service, etc.), looking for ways to ensure that it meets the needs of the user.

The social scientist/researcher serves as the interface between the user and the designer. The researcher collects primary data or uses secondary sources to learn about the needs of the user. The researcher interprets this information, often in the form of design criteria. The designer interprets these criteria, typically through concept sketches or scenarios. The focus continues then on the design development of the thing. The researcher and user may or may not come back into the process for usability testing.

In user-centered design, the roles of the researcher and the designer are distinct, yet interdependent. The user is not really a part of the team, but is spoken for by the researcher.

## Participatory culture

But I can see now, at the end of 1999, that there is a common ground, a new territory being formed by the reciprocal respect between designers and the social scientists. It is clear that social science still has much to offer design, just as design has much to offer the social sciences.

In participatory experiences, the roles of the designer and the researcher blur and the user becomes a critical component of the process. The new rules call for new tools. People want to express themselves and to participate directly and proactively in the design development process.

Today it is not "business as usual" anymore. The rules have changed and continue to change. The new rules are the rules of networks, not hierarchies. People are cynical about the methods and goals of consumerism. The users of products, interfaces, systems, and spaces are realizing that through networking they have an enormous amount of collective influence. They are beginning to use their influence to get what they want, when they want it and how they want it.

## Design for experiencing

Today we are beginning to hear about "Experience Design," whose aim is to design users' experiences of things, events and places. This influence on design can be attributed to a significant literature being written in the social sciences that has begun to acknowledge the role of emotions in human experience (see Jensen 1999, for example).

But we can never really "design experience." Experiencing is a constructive activity. That is, a user's experience (with communication, for example) is constructed of two equal parts: what the communicator provides, and what the communication brings to the interaction. Where the two parts overlap is where the actual communication occurs. Knowing about users' experiences, then, becomes vital to the process of designing the communication. If we have access to both what is being communicated and what experiences are influencing the reception of communication, then we can design for experiencing.

In fact, if we can learn to access people's experiences (past, current and potential), then we can make user experience the source of inspiration and ideation for design. And by making user experience the source of inspiration, we are better able to design for experiencing.

How do we access experience?

There are many ways we can learn from people about their memories, their current experiences and their ideal experiences (Figure 1.1):

- We can listen to what people say
- We can interpret what people express, and make inferences about what they think

- We can watch what people do
- We can observe what people use
- We can uncover what people know
- We can reach toward understanding what people feel
- We can appreciate what people dream.

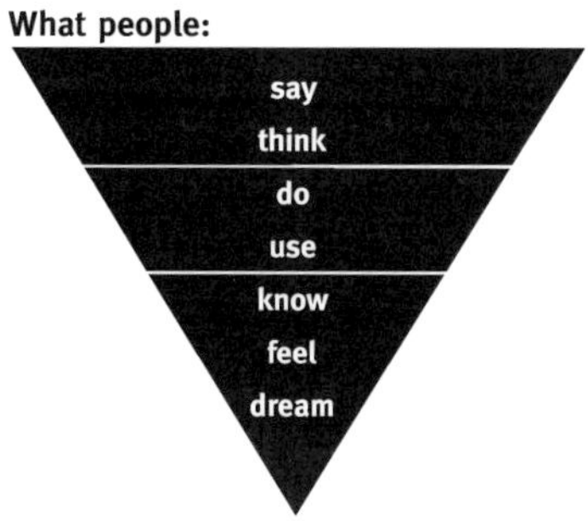


*Figure 1.1* Ways we can learn.

Each route to experience reveals a different story or picture. Listening to what people say tells us what they are able to express in words (i.e. explicit knowledge). But it only gives us what they want us to hear. Watching what people do and seeing what they use provides us with observable information (or observed experience). But knowing what people say/think, do and use (Cain 1998) is not enough (Sanders 1992).

Discovering what people think and know provides us with their perceptions of experience. Understanding how people feel gives us the ability to empathize with them. This way of knowing provides tacit knowledge, i.e. knowledge that can't readily be expressed in words (Polanyi 1983). Seeing and appreciating what people dream shows us how their future could change for the better. It is another form of tacit knowledge that can reveal latent needs, i.e. needs not recognizable until the future (Figure 1.2). For example, the Internet has been revealing many previously latent communication needs.

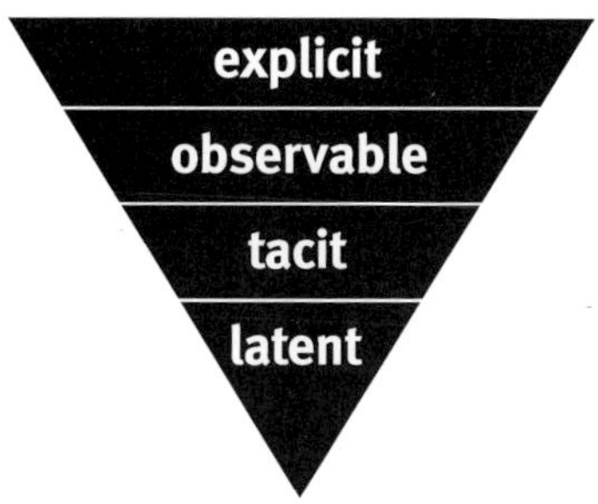


*Figure 1.2* Levels.

The ability to not just know, but also to empathize with the user comes only at the deepest levels of their expression. Special tools are needed to access the deeper levels of user expression. By accessing people's feelings, dreams and imaginations, we can establish resonance with them.

## Accessing experience: what people do, say and make

The different ways of accessing experience have evolved over time. Traditional design research methods were focused primarily on observational research (i.e. looking at what people do and use). Traditional market research methods, on the other hand, have been focused more on what people say and think (through focus groups, interviews and questionnaires). The new tools are focused on what people make, i.e. what they create from the toolkits we provide for them to use in expressing their thoughts, feelings and dreams (Figure 1.3).

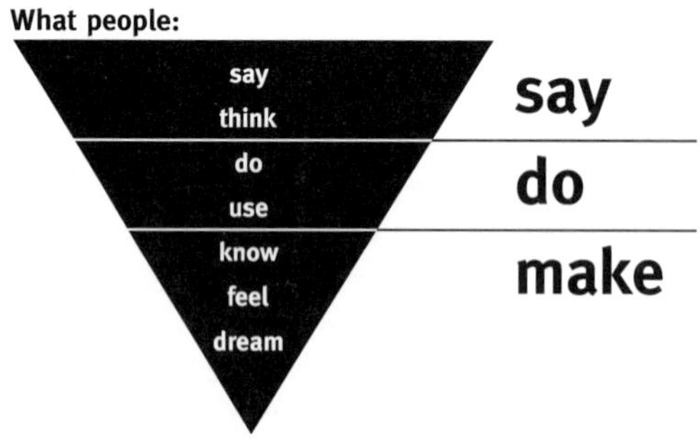


*Figure 1.3* What people say, do, make/1.

When all three perspectives (what people do, what they say, and what they make) are explored simultaneously, one can more readily understand and establish empathy with the people who use products and information systems (Figure 1.4).

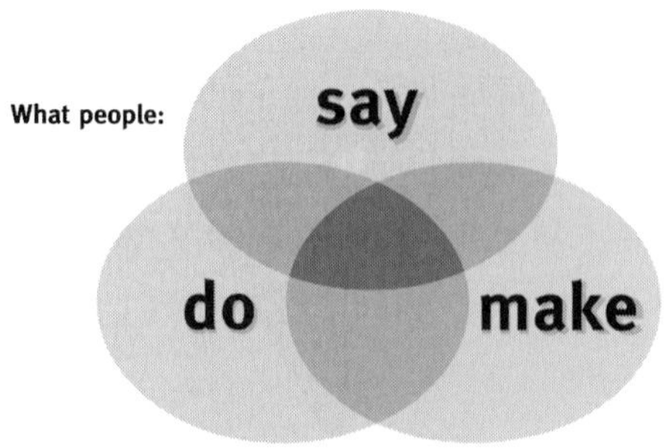


*Figure 1.4* What people say, do, make/2.

## The Make Tools

The Make Tools are the most recent development in design research. Because they are primarily visual, the Make Tools serve as a common ground for connecting the thoughts and ideas of people from different disciplines and perspectives.

The Make Tools are becoming a new language for co-design. They have been found to facilitate exchange between the people who experience products, interfaces, systems and spaces and the people who design for experiencing. The Make Tools are a "design language" for users, not just for designers; a design language built upon an aesthetics of experience rather than an aesthetics of form.

Because they are projective, the Make Tools are particularly good in the generative phase of the design development process. Generative research occurs very early in the design development process. Its purpose is to discover as-yet unknown, undefined and/or unanticipated user or consumer needs. Ideas and opportunities generated by users are usually quite relevant and powerful when acted upon and brought to market.

When Make Tools are used in the generative phase of the design development process, user-generated artifacts result. We have discovered that there are many different types of Make Toolkits that facilitate the expression of a wide range of artifacts and/or models. With "emotional toolkits," people make artifacts such as collages or diaries that show or tell stories and dreams. We have found that these tools are extremely effective in accessing people's unspoken feelings and emotional states. With "cognitive toolkits," people make artifacts such as maps, mappings, 3–D models of functionality, diagrams of relationships, flowcharts of processes and cognitive models.

Every artifact tells a story and so we typically ask the creator of the artifact to tell us that story. The stories associated with the artifacts from the emotional toolkits tell of feelings, dreams, fears and aspirations. The stories associated with the artifacts from the cognitive toolkits tell us how people understand and misunderstand things, events and places. The cognitive toolkits can also reveal the intuitive relationships between system components.

By knowing how to access people's feelings and ideas, we are able to establish resonance between a company and its customers. Resonating, or being in synch with one's customers, means being able to quickly respond to their changing needs and aspirations. Resonance can be achieved by inviting users to play a role in the design development process.

## Collective generativity

We have found that the new tools are effective in accessing end-users' and other people's unspoken feelings and ideas. The ideas they generate tend to be experience-based, not object-based. The tools are projective in nature, allowing users to project their own needs and desires onto their imagined experiences. Artifacts, interfaces, systems and space may or may not play a supporting role in these imaginings. The ideas generated are relevant. Relevance to users means simultaneously useful, usable and desirable.

The new tools can, in fact, harness the collective and infinitely expanding set of ideas and opportunities that emerge when the people who have a stake in the process are invited to "play the game." Generative methods are a new language that enables all the stakeholders to contribute directly to the development of products, goods and services. This new language relies on visual literacy and begins to bring it into balance with verbal literacy.

## Design is changing

How does the emergence of the new tools change the role of the designer? The roles of designer and design researcher are becoming mutually interdependent. The roles are converging to the point where they are blurring. Designers will participate in the creation of the tools and in the expansion of the design language for users. Designers will observe first-hand the experiences the tools afford for creative expression by the users and other stakeholders. Designers will be part of teams responsible for the analysis and interpretation of the "data": the user-generated artifacts and models. Finally, designers can use the ideas generated by the users as sources of design inspiration and innovation.

Who creates the tools for the new design language? Designers and social scientists will need to work together. Social scientists bring frameworks for the understanding of user experience to the table, while designers know how to synthesize and embody ideas and opportunities.

How does the emergence of the new tools change the nature of design education? Designers need to be trained to go beyond the individualized expression of visual communication. They need to learn how to become involved in the creation and construction of the new tools.

Where does Postdesign fit? Postdesign is a new mindset. It transcends the traditional domain of design by making user experience (as opposed to artifacts, interfaces, systems or spaces) the focus for design inspiration and ideation. It is easy to see that people are ready for the Postdesign mindset. Just look at the Internet. New computer tools and applications have made self-expression through personal websites accessible to everyone with the time and desire to build one.

Postdesign is not about specific methods, tools or processes. It is about an emerging visual language that people, all people, can use to express and interpret those ideas and feelings that are often so difficult to express in words.

Postdesign is an attitude about people. It is about the recognition that all people have something to offer and that they, when given the means to express themselves, can be both articulate and creative.

Postdesign is contextual. Understanding and empathizing with the people who experience artifacts, interfaces, systems and spaces can best be accomplished by communicating with them in the places where they live, work and play while they live, work and play.

Postdesign is participatory. It emphasizes the direct and active participation of all stakeholders in the design development process. This makes the deliverables of design more meaningful to the people who will ultimately benefit from them.

Postdesign is co-design, i.e. people designing together. It can harness the collective and infinitely expanding set of ideas and opportunities that emerge when all the people who have a stake in the process are invited to "play the game."

Postdesign is an ongoing process. People's needs change and their experiences change. Relationships between people change over time, as well. Postdesign is not a linear process but a continual intersection of changing perspectives. Today it blends design and the arts with the applied social sciences and blends them both with new and emerging technologies.

The challenge ahead for the Postdesign community is to create the tools and infrastructure needed to support and to facilitate continued resonance with user experience.

## Bibliography

Barlow, J. P. (1994) "The economy of ideas." *Wired*, March, 85.

Cain, J. (1998) "Experience-based design: toward a science of artful business innovation." *Design Management Journal*, Fall.

Dandavante, U., Sanders, E. B.-N. and Stuart, S. (1996) "Emotions matter: user empathy in the product development process." In *Proceedings of the Human Factors and Ergonomics Society 40th Annual Meeting* (Santa Monica: Human Factors and Ergonomics Society), 415–18.

Jensen, R. (1999) *The Dream Society: How the Coming Shift from Information to Imagination will Transform your Business* (New York: McGraw-Hill).

Kelly, K. (1998) *New Rules for the New Economy* (New York: Viking).

Krippendorff, K. (1996) "A trajectory of artificiality and new principles of design." Unpublished manuscript.

Leinbach, C.O. (1999) "About cake and the top ten things (more or less) Dorothy must have been thinking when she said, Toto, I've a feeling we're not industrial anymore." Unpublished manuscript.

Pine, B.J. and Gilmore, J.H. (1999) *The Experience Economy: Work is Theatre and Every Business a Stage* (Boston: Harvard Business School Press).

Polanyi, M. (1983) *The Tacit Dimension* (Gloucester, Mass.: Peter Smith).

Postrel, V. (1998) *The Future and its Enemies: The Growing Conflict over Creativity, Enterprise, and Progress* (New York: The Free Press).

Reed, E. S. (1996) *The Necessity of Experience* (New Haven: Yale University Press).

Sanders, E.B.-N. (1992) "Converging perspectives: product development research for the 1990s." *Design Management Journal*, Fall.

Sanders, E.B.-N. (1994) "But is it useful? Testing beyond usability." *Innovation*, Spring.

Sanders, L. (1997) *Cognition and Emotion Bibliography, A Project of the Graphic Design Education Association*, American Center for Design, Chicago, Illinois.

Sanders, L., Dandavate, U., and Stuart, S. (1996) "User empathy in the product development process." *Proceedings of IDSA Conference on Alternate Realities* (on CD-ROM), IDSA.

Schlossberg, E. (1998) *Interactive Excellence, Defining and Developing New Standards for the Twenty-First Century* (New York: The Ballantine Publishing Group).

Wechsler J. (ed.) (1988) *On Aesthetics in Science* (Boston and Basel: Birkhauser).

# 2 Human factors for pleasure seekers

*Patrick W. Jordan*

## Introduction

Humans always have and always will seek pleasure. The artifacts and products with which we surround ourselves are potential sources of pleasure. The role and methodologies of the human factors profession make the discipline the natural vehicle for assuring that products are designed such that they are pleasurable for those who use and experience them.

This chapter describes and defines the concept of pleasure with products. This is explained in the context of a hierarchy of user needs. A framework in which to consider pleasure issues is given and the challenges that human factors faces in order to assure product pleasurability are discussed.

## In search of pleasure

Since the beginning of time humans have sought pleasure. We have gained pleasure from the natural environment. From the beauty of flowers or the feeling of the sun on our skin. From bathing in soothing waters or the refreshment of a cool breeze. We have actively sought pleasure, creating activities and pastimes to stretch our mental and physical capabilities or to express our creative capabilities. Cave-dwellers wrestled to test their strength and expressed themselves through painting on the walls of their dwellings. Today we "pump iron" in the gymnasium and decorate our homes with selections of paintings and posters.

Another source of pleasure has been the artifacts with which we have surrounded ourselves. For centuries humans have sought to create functional and decorative artifacts. Artifacts that have increased the quality of life and brought pleasure to the owners and users. Originally, these objects would have been clumsily bashed out from stone, bronze or iron by unskilled people who simply wanted to make something for their own use. As systems of trade and barter were developed specialist craftspeople became prevalent, creating artifacts for use by others in the community. Today, most of the artifacts that we surround ourselves with were created by industry.

## Design and human factors in product creation

The product creation process will differ from product to product. However, it is probably a truism that design is integral to the creation of any product. Whereas the craftsperson would have been both designer and manufacturer of a product, the scale and economics of industrial mass production have led to an increasing specialization of roles within industry. One of these roles is, of course, that of professional designer.

Nowadays, professional designers will have been involved in the creation of virtually all the products that we find in our homes, communities or workplaces. Much of industry—especially those sectors of industry connected with the creation of consumer products—has come to recognize the need to produce designs that are well thought out with respect to their fit to the user. This has been good news for human factors specialists. Increasingly, human factors specialists have been employed to advise designers as to how best to match a product to user needs. Indeed, the number of human factors specialists employed in industry is now at an unprecedented high.

## Usability

Usually, the approach taken by human factors specialists towards fitting the product to the person has been to concentrate on the issue of usability. In layperson's terms this means ensuring that the product is easy to use. More formally, the International Standards Organization (ISO) defines usability as: "the effectiveness, efficiency and satisfaction with which specified users achieve specified goals in particular environments" (ISO DIS 9241–11).

Effectiveness refers to the extent to which a goal, or task, is achieved; efficiency, to the amount of effort required to accomplish a goal and satisfaction to the level of comfort that the user feels when using a product, and how acceptable the product is to users as a means of achieving their goals. In effect, then, a usability-based approach to user-centered design is one which sees the product as a tool with which users try to accomplish particular tasks without wanting to have to expend unnecessary effort or endure any physical or mental discomfort. This definition has received wide acceptance as a basis for much of the human factors work carried out in industry.

Human factors has become very adept at assuring usable design. For example, there has been a lot of work done to establish the particular properties of a design that will affect usability. Ravden and Johnson (1989), for example, link usability with the design properties such as: consistency, compatibility, feedback, visual clarity and error prevention and recovery.

Similarly, a battery of methodologies for evaluating usability has been established. The majority of these were originally developed in psychology and have been adapted specially for the evaluation of product usability. For example, the video "User Interface Performance Measurement" includes demonstrations of focus groups, incident diaries, questionnaires, interviews, think-aloud protocols, feature checklists and experiments. Jordan (1993) gives brief written descriptions of the same methods.

## Why human factors has been successful

As a discipline then, human factors is well equipped to assure the usability of products. This has been the cornerstone of the profession's success in an era when consumers are no longer willing to accept lack of usability as a price to pay for "technical wizardry." People are becoming disenchanted with difficult products. In the end, of course, these products will also disenchant those who manufacture them as they will find that their customers start looking elsewhere. Indeed, the issue of good design as a whole is moving nearer and nearer to the top of manufacturers' agendas. This is largely due to the "technology ceiling" that has been reached in many sectors of manufacturing industry. With technology being so advanced now, it is difficult for one manufacturer to gain a significant advantage over another in terms of, for example, production cost or technical reliability (Jordan *et al.* 1996). Design, then, is one of the few areas where there can now be real competition. Usability is becoming recognized by manufacturers and consumers alike as a prerequisite of good design. More and more products are being advertised as "user friendly" or "ergonomically designed" and customers increasingly consider usability as a criterion when making purchase choices.

## Usability engineering as dehumanization

Usability is an important issue and usability based approaches have undoubtedly brought huge benefits to users of products. After all, what is the point of providing users with vast arrays of functions if the design of the product makes it difficult to use them to their full advantage? Nevertheless, usability-based approaches are inherently limited.

The reason why they are limited is that usability-based approaches tend to look at products as tools with which users complete tasks. However, products are not merely tools. Products are living objects with which people have relationships. Products are objects which can make people happy or angry, proud or ashamed, secure or anxious. Products can empower, infuriate, delight—they have personality (Jordan 1997).

People also have personalities. Not only do they have personalities, they have hopes, fears, dreams and aspirations. These are liable to affect the way that people respond to and interact with products. Again, this may seem obvious at first. However, if a being from another planet were to try to learn about the human race via the human factors literature, he/she/it would probably conclude that we were basically little more than cognitive and physical processors. It is a rarity to find published human factors studies that describe users in terms that go beyond factors such as age, sex, education or profession. How much work has human factors done on matching products to people's personalities, their emotional responses, or their ideals? Very little! (Although see Dandavante *et al.* 1996, and Rijken and Mulder 1996, for rare—and good—examples.)

This, then, is the problem with usability-based approaches—they tend to encourage the view that users are merely cognitive and physical components of a system comprising of user/system/environment. The idea is that the product must be designed such that the demands it places on this cognitive and physical component are minimized.

It is important to recognize and emphasize that usability is a very important issue. However, there must also be an awareness that usability is only one of the issues that will affect the overall relationship between a person and a product. The problem with usability based approaches is that they encourage a limited view of the person using the product. This is—by implication if not by intention—dehumanizing.

This seems ironic—of all the people involved in the product creation process, it is the human factors specialist, with his or her roots in behavioral science, who is the person that would be expected to have the richest understanding of users. It is no longer sufficient for the profession to think of users in such limited terms. In order to represent the user fully in the product creation process, human factors specialists must take a wider view of person-centered design and look both at product use and at those using and experiencing products in a more holistic context.

## Human factors and pleasure

The rest of this article outlines a pleasure-based approach to person-centered design. This is an approach which seems to offer the scope for human factors to broaden and extend its influence on the product creation process, to move products beyond being usable to the stage where they are not only usable but also enjoyable, exciting and meaningful— pleasurable.

Pleasure: What is pleasure? The Oxford English Dictionary defines it as "the condition of consciousness or sensation induced by the enjoyment or anticipation of what is felt or viewed as good or desirable; enjoyment delight, gratification. The opposite of pain."

In the context of products, Jordan (1996) defines pleasure as "the emotional and hedonic benefits associated with products use." That definition now seems somewhat problematic as it seems to ignore the contribution to pleasure of a product's practical benefits—making it open to the opposite charge of negligence that has here been leveled at usability-based approaches! A more complete definition would be:

Pleasure with products: the emotional, hedonic and practical benefits associated with products.

## Hierarchy of user needs—functionality, usability and pleasure

In the 1996 keynote address to the European Conference of the Human Factors and Ergonomics Society (Jordan 1999) it was argued that looking at pleasure with products is a natural progression from usability-based approaches. An analogy was drawn with Maslow's hierarchy of needs.

Abraham Maslow (1970) developed a "hierarchy of human needs." This model views the human as a "wanting animal" who rarely reaches a state of complete satisfaction. Indeed, if a nirvana is reached it will usually only be temporary because once one desire has been fulfilled another will soon surface to take its place. The idea is that as soon as people have fulfilled the needs lower down the hierarchy, they will then want to fulfill the needs higher up. This means that even if basic needs such as physiological needs and safety have been met, people will still meet with frustration if their higher goals are not met. (A good overview of Maslow's work can be found in Hjelle and Ziegler 1981.)

Taking the idea of a hierarchy of needs and applying it to human factors, Jordan (1999) developed the following hierarchy of user needs: (1) functionality; (2) usability; (3) pleasure.

### *Level 1—functionality*

Clearly a product will be useless if it does not contain appropriate functionality. A product cannot be usable if it does not contain the functions necessary to perform the tasks for which it is intended. If a product does not have the right functionality it will dissatisfy the user. In order to be able to fulfill user needs on this level, the human factors specialist must have an understanding of what the product will be used for and the context and environment in which it will be used.

### *Level 2—usability*

Once users have gotten used to having appropriate functionality, they will then want products that are easy to use. This more or less represents the situation at the moment in many product areas—people are used to well-functioning products, now they expect usability too. Having appropriate functionality is a prerequisite of usability, but it does not guarantee usability. To assure usability the human factors specialist must have an understanding of some or all of the design principles discussed by Ravden and Johnson (1989) and the methods described by Jordan (1993).

### *Level 3—pleasure*

Having gotten used to usable products, it seems inevitable that users will soon want something more. Products that offer something extra. Products that are not merely tools, but which are "living objects" which people can relate to. Products that bring not only functional benefits but also emotional benefits. To achieve product pleasurability is the new challenge for human factors. It is a challenge that requires an understanding of people—not just as physical and cognitive processors—but as rational and emotional beings with values, tastes, hopes and fears. It is a challenge that requires an understanding of how people relate to products. What are the properties of a product that elicit particular emotional responses in a person. How does a product design convey a particular set of values? Finally, it is a challenge that requires capturing the ephemeral—devising methods and metrics for investigating and quantifying emotional responses.

## The four pleasures—a new framework for human factors

So, a hierarchy of user needs has been given and a definition of pleasure with products has been offered as well as a dictionary definition of pleasure. But none of this really gives much of a feel for what pleasure is or how it could be applied in the context of products.

Unfortunately, the human factors literature is not helpful. This is not surprising—after all one premise on which the arguments presented here are based is that pleasure is an issue that has been largely ignored by the human factors profession. In the absence of human factors literature on the issue, the most appropriate starting point would seem to be literature from the behavior sciences.

A useful way of classifying different types of pleasure has been espoused by Canadian anthropologist Lionel Tiger. Tiger has made a study of pleasure and has developed a framework for addressing pleasure issues (Tiger 1992). The framework models four conceptually distinct types of pleasure—physical, social, psychological and ideological. Summaries of Tiger's descriptions of each are given below. Examples are added to demonstrate how each of these components might be relevant in the context of products.

### *Physio-pleasure*

This is to do with the body—pleasures derived from the sensory organs. They include pleasures connected with touch, taste and smell as well as feelings of sexual and sensual pleasure. In the context of products physio-pleasure would cover, for example, tactile and olfactory properties. Tactile pleasures concern holding and touching a product during interaction. This might be relevant, for example, in the context of a telephone handset or a remote control. Olfactory pleasures concern the smell of the new product. For example, the smell inside a new car may be a factor that affects how pleasurable it is for the owner.

### *Socio-pleasure*

This is the enjoyment derived from the company of others. For example, having a conversation or being part of a crowd at a public event. Products can facilitate social interaction in a number of ways. For example, a coffee-maker provides a service which can act as a focal point for a little social gathering—a "coffee morning." Part of the pleasure of hosting a coffee morning may come from the efficient provision of well-brewed coffee to the guests.

Other products may facilitate social interaction by being talking points in themselves. For example a special piece of jewelry may attract comment, as may an interesting household product, such as an unusually styled TV set. Association with other types of products may indicate belonging in a social group—Porsches for "Yuppies," Dr. Martin's boots for skinheads. Here, the person's relationship with the product forms part of their social identity.

### *Psycho-pleasure*

Tiger defines this type of pleasure as that which is gained from accomplishing a task. It is the type of pleasure that traditional usability approaches are perhaps best suited to addressing. In the context of products, psycho-pleasure relates to the extent to which a product can help in accomplishing a task and make the accomplishment of that task a satisfying and pleasurable experience. For example, it might be expected that a word processor which facilitated quick and easy accomplishment of, say, formatting tasks would provide a higher level of psycho-pleasure than one with which the user was likely to make many errors.

### *Ideo-pleasure*

Ideo-pleasure refers to the pleasures derived from "theoretical" entities such as books, music and art. In the context of products it would relate to, for example, the aesthetics of a product and the values that a product embodies. For example, a product made from bio-degradable materials might be seen as embodying the value of environmental responsibility. This, then, would be a potential source of ideo-pleasure to those who are particularly concerned about environmental issues. Ideo-pleasure would also cover the idea of products as artforms. For example, the video cassette player that someone has in the home, is not only a functional item, but something that the owner and others will see every time that they enter the room. The level of pleasure given by the VCR may, then, be highly dependent on how it affects its environment aesthetically.

### *Challenges for human factors*

Having a hierarchy of user needs, a definition of pleasure with products and now a framework for looking at different types of pleasure, seems a constructive basis from which human factors can move towards tackling pleasure issues. However, this basis does not of itself indicate how human factors as a profession can contribute to the creation of pleasurable products. In order to achieve this, there are at least three issues that the profession must tackle. These are understanding users and their requirements, linking product properties to emotional responses in order to fulfill these requirements, and developing methods for the investigation and quantification of pleasure.

## Understanding users and their requirements

Earlier, it was asserted that usability-based approaches tend to encourage human factors specialists to consider people as processors. Physical processors with attributes such as strength, height and weight, and cognitive processors with attributes such as memory, attention and expectations. Here, then, the user is often looked at as being simply a cognitive and/or physical component of a three component system—the other two components being the product and the environment. It could be argued that the traditional human factors approaches to people ignore the very things that make us human—our emotions, our values, our hopes and our fears.

In order to find a way into these issues, we need to have an understanding not only of how people use products, but also of the role that those products play in people's lives. This gives a chance to understand how the product relates to the person in a wider sense than just usability and can help the human factors specialist in gaining a wider view of the user requirements—the requirements for pleasure.

As an illustration of the pleasure benefits that products can bring to their users, three case studies are given below. These case studies arose from interviews conducted for the pleasure study reported by Jordan and Servaes (1995). They are also included in Jordan (1999).

## Case studies

### *Case study 1: the hairdryer user*

This 17-year-old woman chose a hairdryer as her pleasurable product. A product which she described as being "perfect . . . the best (hairdryer) I've ever had." The reason she was so positive about the hairdryer was because it helped her to style her hair in just the way she wanted. This made her feel attractive and gave her a feeling of self-confidence when she went out. She also mentioned that the hairdryer had an unusual design and was thus something that caught the attention of her friends, "it's 'showy,' I like it when people come into my room and see it."

Both of the pleasures she mentions fall in the socio-pleasure part of Tiger's framework. They are both concerned with the enhancing the image of herself that she projects to others (somebody good-looking, somebody with interesting tastes) and how she feels in the company of others (self-confident).

### *Case study 2: the guitar player*

The guitar player was a 26-year-old man who played the electric guitar. Again, the guitar facilitated socio-pleasure. He regarded it as a "status symbol particularly amongst people who know about these things." It had also provided a talking point as it had belonged to Lloyd Cole—lead singer of Lloyd Cole and the Commotions, a Glaswegian band who had a number of hits in the 1980s including the single "Brand New Friend."

He also found playing the guitar an exciting activity in itself. This would probably classify as a psycho-pleasure or possibly an ideo-pleasure (all four categories can potentially overlap). Even just having the guitar near him gave him a feeling of re-assurance, again either an ideo- or psycho-pleasure.

### *Case study 3: the video cassette recorder (VCR) user*

She was a 26-year-old living alone who rented a VCR which she described as being a "standard video." She described the emotional benefits that she gained from the video as being a feeling of anticipation—looking forward to watching what she had recorded (an ideo-pleasure) and freedom—not having to stay home to catch her favorite programs (a socio-pleasure as she could go out with her friends).

So, even in this little selection, the benefits that people mentioned went well beyond being merely comfortable and acceptable—the satisfaction component of usability. Ensuring that these products were effective, efficient and satisfying to use would stop well short of providing the benefits mentioned by these three people. It is clear, for example, that in case studies 1 and 2, the person/product relationship went well beyond that of user/tool. The hairdryer was an "object d'art," the guitar was a status symbol, a talking point, even an old friend.

The first challenge, then, is to investigate and inventorize the types of pleasure that products can potentially bring to their users based on a holistic understanding of the role which the product plays in a person's life.

It should be fairly straightforward to move forward here. In the first instance it may simply be a matter of asking the appropriate questions during evaluations and requirements capture sessions. Not simply asking about functional benefits but also about the types of pleasure that a product can bring. Certainly, pleasure is not a straightforward issue: people may not always be able or willing to articulate descriptions of the types of pleasure that they gain from a particular product. And certainly what people do say may often be difficult to interpret. Nevertheless, simply having an awareness of pleasure issues and asking some sensible questions will very quickly move human factors a long way towards meeting this challenge.

## Linking product properties to pleasure benefits

Having established the different types of pleasure that people can get from products, the next stage is to link those to particular product properties. For example, it might be that feelings of security with a product are linked to high levels of usability and/or high levels of product reliability. Similarly, a feeling of pride may be linked to, say, good aesthetics. Similarly, particular types of displeasure may also be linked to inadequacies with respect to certain product properties. For example, annoyance might be linked to poor technical performance whilst anxiety may be related to a lack of usability.

All the above suggestions are, of course, merely speculations. This is an issue that human factors must address systematically if it is to make a significant contribution to the development of pleasurable products. One way in which to approach this would be to correlate people's pleasure responses to a product to the "goodness" of the product with respect to various properties. This type of approach has been taken within Philips Design in the context of analyzing the requirements for audio systems. Unfortunately, the outcomes of the study are still commercially confidential, however the approach is described briefly here.

Interviews were conducted with eighty stereo owners. They were asked to rate their stereos with respect to various properties such as usability, sound quality, aesthetics and power. They did this by marking Lickert scales. They were then asked to mark a second set of Lickert scales indicating their emotional responses to the product. This was done through marking their levels of agreement or disagreement with statements such as "This stereo gives me a sense of pride" or "I feel anxious with this stereo." It was then possible to correlate the ratings of the stereos properties with the ratings on the emotional dimensions in order to link particular properties with particular emotions.

Admittedly, this study was, in the context of the constraints usually imposed on industry-based human factors work, a very thorough and costly approach to this issue. However, this sort of approach is certainly one that could be followed by those in research and academia who are supporting industry-based human factors specialists. A less formal approach is the case study. Discussing users' experience of a product, what they perceive as that product's properties and the types of pleasurable emotions that they feel with the product. Below are some examples, again drawn from the interviews reported in Jordan and Servaes (1995).

### *Case study 4: the CD player owner*

This 19-year-old man had received a stereo as a Christmas present. When asked about what aspects of the product he found particularly appealing he mentioned that the CD player was "less hassle" than a tape player—he had previously owned a tape player—as the CD would play for longer than a tape without him having to turn it over. It was also possible to put the CD in repeat play mode, which again enabled him to play music for a long period without having to interact with the player at all. When asked about the emotional benefits, he said that the product gave him a sense of freedom as he was able to set it playing and then listen to the music whilst doing something else. Here, then, there was an association between one specific feature—repeat play—and one particular emotional benefit—a sense of freedom.

### *Case study 5: another hairdryer owner*

This 19-year-old woman had been given the hairdryer as a present 11 years ago. She found the product particularly pleasurable for two reasons. Like the first hairdryer user she said that she could use the dryer to style her hair in the way that she wanted and thus it gave her a feeling of confidence in her appearance. Unfortunately, it was not clear which property of the dryer had particularly contributed to making it so suitable for styling—presumably it was some combination of technical performance, functionality and usability—but this wasn't expressed clearly.

In addition, though, she also noted that the product's reliability gave her a feeling of confidence in the product itself. Here, then, there is a link between the property of product reliability and the emotional response of confidence in the product.

### *Case study 6: the TV watcher*

This 19-year-old woman chose her parents' television as her pleasurable product. She described this as being big and straightforward with an easy to use remote control. She said that she particularly liked the TV because of its simplicity, because of the large screen and the good picture quality, and because of the wide variety of programs that she could choose from (her parents had a cable TV subscription). Again, though, she didn't link these qualities to any particular emotional response, other than indicating that, taken together, they led to her feeling "satisfied" with the TV.

In addition, however, she mentioned that the TV was very reliable and that this enabled her to take the TV for granted. Here, then, there was a link between the property of reliability and a feeling of "security" that nothing would go wrong.

## Developing methods and metrics for the investigation and quantification of pleasure

A major aid to the incorporation of usability issues in the design process has been the ability to quantify these issues. In the case of effectiveness and efficiency this has usually meant taking performance measures of a user with a product. These include, for example, task success and quality of output (effectiveness) and time on task and error rate (efficiency). Quantitative attitude scales, such as the System Usability Scale (Brooke 1996) and the Software Usability Measurement Inventory (Kirakowski 1996), have also been developed to measure the satisfaction component of usability.

Being able to quantify usability has enabled human factors specialists to set usability specifications which have then been included as part of the overall product specification. As well as giving clear usability targets to aim at, this approach has proved effective as it has given clear signals to others involved in the product creation process as to what is meant by usability in a particular context. Talking in terms of "usability" or "quality of use" can seem vague and "woolly," particularly to technical colleagues, who are used to numerical specification. Quantifying the issues gives an unequivocal signal as to what is required and enables an equally firm judgement as to whether or not the criteria have been met. It follows, then, that if human factors as a profession is to take the lead in ensuring product pleasurability, then the development of measures and tools for quantifying pleasure will be beneficial if not essential.

Again, one approach to this would be to develop attitude scales for measuring pleasure. This requires a knowledge not only of the potentially pleasurable benefits that could be associated with product use, but also an idea of the comparative importance of each, and the relationship between the benefits—which benefits are conceptually separate in the context of associations with products (i.e. which pleasure benefits are potentially, if not always, independent from each other).

Another way to investigate pleasure with product use may be to look at potential behavioral correlates to pleasure and displeasure. For example, the frequency with which a person smiles when using a product may be seen as a simple measure of pleasure. Similarly, frowning may be seen as a simple measure of displeasure. Whilst measures such as these may be simple to take and appear to have a degree of objectivity that is lacking in questionnaire- or interview-based approaches, they still appear, a priori, to have a number of drawbacks associated with them. For example, there are some pleasure benefits with which smiling may not be associated—many of the ideo-pleasures for example. Secondly, there are many reasons for which a person may smile that have nothing to do with what is being experienced with the product. Maybe they are thinking about something else, maybe they are amused at how awful the product is!

Despite these drawbacks, facial expressions do seem a promising way forward. Facial expressions have been used as an investigation method in psychology and there is a precedent for their use in human factors in the domain of human–computer interaction (HCI). For an example of the former see Ekman and Friesen's (1978) proposed emotion coding system, whilst Oatley and Ramsay (1992), provide a good example in the context of HCI.

Other simple metrics of pleasure may be the frequency with which people use a particular product—where use of the product is voluntary—and hopefully, from the point of view of those who manufacture pleasurable products, purchase choice.

Of course, creating pleasurable products requires more that just effective evaluation methods. It is also very important to have methods available for capturing user requirements with respect to pleasure as a starting point in the product creation process. Similarly, methods for early concept evaluation are also needed.

Human factors already has many methods that are likely to be effective here. For example, the interview, focus group and questionnaires would surely be staples for anyone addressing these issues. Nevertheless, it would be prudent to look at other potential methods for addressing these issues. One promising new method is the Private Camera Conversation, developed by de Vries, Hartevelt and Oosterholt (1996). This involves participants giving monologues to a camera in response to written questions. De Vries *et al.* claim that this method can provide rich information about users relationships with products—participants cover matters that they may seem reticent to talk about face to face with an investigator for fear that it may appear foolish to relate to a product emotionally.

Another potentially promising method is the repertory grid (Kelly 1955). Baber (1996) suggests that the approach would be useful as a means of defining users' conceptions of usability. Presumably, the same could hold true for investigating pleasure with products.

## Final thoughts

People always have and always will seek pleasure. The artifacts and products with which we surround ourselves are potential sources of pleasure. They should be designed with a view to how they can provide pleasure to those who use and experience them. Human factors, as a profession, is in a unique position to be able to influence the creation of pleasurable designs.

Human factors has become very adept in linking design decisions with usability and, thanks to the development of a variety of methods, in the investigation of usability issues. Similar approaches can be taken with respect to product pleasurability. However, in order to be able to address pleasure issues effectively, three major challenges must be met:

1 Understanding users and their requirements—understanding the emotional and hedonic benefits that a product can bring to users.
2 Linking product properties to pleasure benefits—which properties of a product are associated with which benefits.
3 Developing methods and metrics for the investigation and quantification of pleasure—the ability to quantify usability has been a central reason for the ever-growing influence of human factors on the product creation process. A similar approach is required if colleagues in other disciplines are to be persuaded to take pleasure seriously as an issue.

Over the last two or three decades, human factors as a profession has achieved a great deal in terms of enhancing the usability of products. The profession can now move a stage further. By moving beyond usability to take a holistic pleasure-based approach, the profession can facilitate the creation of useful, usable and pleasurable products that will delight those who use and experience them.

## Note

1 This paper is based on the presentation "Ergonomics for Pleasure Seekers." This presentation was given by the author to press and public at the Milan Furniture Fair on 12 April 1997 at the request and invitation of Assarredo/Federlegno-Arredo/Società Italiana di Ergonomia.

## References

Baber, C. (1996) "Repertory grid theory and its application to product evaluation." In Jordan, P.W. *et al.* (eds) *Usability Evaluation in Industry* (London: Taylor and Francis).

Brooke J. (1996) "SUS—a quick and dirty usability scale." In Jordan, P.W., *et al.* (eds) *Usability Evaluation in Industry* (London: Taylor and Francis).

Dandavante, U., Sanders, E. B.-N. and Stuart, S. (1996) "Emotions matter: user empathy in the product development process." In *Proceedings of the Human Factors and Ergonomics Society 40th Annual Meeting* (Santa Monica: Human Factors and Ergonomics Society), 415–18.

Ekman, P. and Friesen, W. V. (1978) *Facial Action Coding System* (Palo Alto, Calif.: Consulting Psychologists Press).

Hjelle, L. A. and Ziegler, D. J. (1981) *Personality Theories* (London: McGraw-Hill).

ISO DIS 9241–11. *Ergonomic Requirements for Office Work with Visual Display Terminals* (VDTs). Part 11: "Guidance on Usability" (Geneva: ISO).

Jordan, P. W. (1993) "Methods for user interface performance measurement." In Lovesey, E.J., (ed.) *Contemporary Ergonomics* (London: Taylor and Francis).

Jordan, P.W. (1996) "Displeasure and how to avoid it." In Robertson, S. (ed.) *Contemporary Ergonomics* (London: Taylor and Francis).

Jordan, P.W. (1997) "Products as personalities." In Robertson, S., (ed.) *Contemporary Ergonomics* (London: Taylor and Francis).

Jordan, P.W. (1999) "Pleasure with products: human factors for body, mind and soul." In Green, W.S. and Jordan, P.W., (eds) *Human Factors in Product Design: Current Practice and Future Trends* (London: Taylor and Francis).

Jordan, P. W., McClelland, I. L. and Thomas, B. (1996) "Introduction." In Jordan, P.W. *et al.* (eds) *Usability Evaluation in Industry* (London: Taylor and Francis).

Jordan, P. W. and Servaes, M. (1995) "Pleasure in product use: beyond usability." In Robertson, S. (ed.) *Contemporary Ergonomics* (London: Taylor and Francis), 341–6.

Kelly, G. A. (1955) *The Psychology of Personal Constructs* (New York: Norton).

Kirakowski, J. (1996) "The software usability measurement inventory: background and usage." In Jordan, P.W. *et al.*, (eds) *Usability Evaluation in Industry* (London: Taylor and Francis).

Maslow A. (1970) *Motivation and Personality* (2nd edn) (New York: Harper and Row).

Oatley, K. and Ramsay, J. (1992) "Emotions while interacting with computers." *Proceedings of The International Society For Research on the Emotions Annual Conference.* Carnegie Mellon University, USA (Storrs, Conn.: ISRE).

Ravden S.J., and Johnson G.I. (1989) *Evaluating Usability of Human-Computer Interfaces: a Practical Method* (Chichester: Ellis Horwood).

Rijken, D. and Mulder, B. (1996) "Information ecologies, experience and ergonomics." In Jordan, P.W., *et al.*, (eds) *Usability Evaluation in Industry* (London: Taylor and Francis).

Tiger, L. (1992) *The Pursuit of Pleasure* (Boston: Little, Brown and Company), 52–60.

Vries, G. de, Hartevelt, M. and Oosterholt, R. (1996) "Private Camera Conversation method." In Jordan, P.W., *et al.*, (eds) *Usability Evaluation in Industry* (London: Taylor and Francis).

# 3 Communications artifacts

## The design of objects and the object of design

*Jim Wilson*

### Making successful products and services

Product development is a balancing act of competing, and sometimes contradictory, objectives. Over 10,000 new products are marketed in the US every year (Crawford 1994). For those which reach the market (i.e. not including those canceled prior to launch), roughly 30 to 40 percent fail—in fact, some cite figures as high as 80 percent. This failure rate has not changed much in the past 15 years. What is failure? Failure is not achieving success! Product success is typically defined as including one or more of the following: achieving the expected profit; achieving the expected market sales or share; meeting the required quality standards; delivering on the specified launch date; achieving user acceptance; meeting the requirements of the customer.

The present chapter provides just one industrial perspective on "Design and the Social Sciences." It will touch on just two of the key components: "usability" and "design." The objective of the conference was to "identify and define possible directions for interdisciplinary work connecting design and the social sciences regarding the conception, production and use of objects and communications." In fact, the present chapter will attempt to describe the "the conception, production and use of objects and communications" in the conception, production and use of the current and future objects for "communication" itself.

My wish list: what I look for in potential student employees, and what I too often find. Looking back at my 5 September notes from the 1996 Human Factors and Ergonomics Society Annual Meeting. I have written: "We aren't turning out human factors students who are interested and equipped to do product design."

My experience in interviewing potential job candidates at the conference (it seems to reflect my most common time at conferences) led to my frustration. It is not that I am saying that a university should be just a vocational training school, but to meet my needs (and those of many others) it must foster in student development the recognition of needs that are not just academic. Certainly, the nature of this process should depend on the current student level of graduation (e.g., undergraduate, masters and doctorate).

What I wish for in a potential employee is someone who: has excellent communication skills; is team-oriented; thrives on changing demands; knows when to fight (and when to stay quiet); has the ability to learn and reflect; and has a passionate (but not strident) desire to develop products. Unfortunately, what I too often find is students who have been promoted and rewarded to be unique, private and competitive individuals who work against, rather than with, their co-employees.

I wish that students received a more balanced and realistic idea of the roles of human factors in product development—a more balanced exposure to the various roles one plays as a product development specialist—one that shows that it is not just "usability testing."

I wish that students had the ability to do excellent research—concept, design, execution, analysis, interpretation, and publication—but also had the ability to know that most of the time this is not needed.

I wish they realized that their education is only a foot in the door. Industry is not looking to fill an academic position—so I do not need to see their schools listed as first on their resume. Worst of all, the student should not tell me that they would take a job in either academia or industry—they won't give me a warm feeling about their passion for product design.

Finally, I wish that faculty advisors were not the ones who encouraged the academic-centered resume. Many students spend their entire academic career being "trained" to do a job that their trainers truly do not understand. Product development is not like doing academic research, it is not like doing government sponsored research, for the most part, it is not like doing research at all!

## Can "design and the social sciences" really make a difference in product and service success?

Product success depends on building the right product, in the right way, at the right price, and on time. The way consumers perceive the relationship between physical characteristics and product consequences is often more important than the physical product characteristics themselves. The application of the social sciences can help to further explain and promote the benefits of new product design to effectively meet these needs. Benefits can be seen as the (perceived) advantages from using a product—people often seek the benefits (including the changes in their emotions, their status, etc.) that a product delivers rather than the product per se.

Confronted with a radically new product, consumers are generally not able to link physical product characteristics with the consequences (advantages, benefits, disadvantages, total cost of ownership, etc.) of owning and using the product. The way consumers perceive the relationship between physical characteristics and product consequences may be more important than the physical product characteristics themselves.

Directly asking customers or users about their needs and wants will typically lead to a narrow focus on their solutions to their perceived problems. To design and sell effective new products those involved may want to understand potential customer problems in their context of use, not just what they think they need. To really succeed, you must ensure that the customer's total product interaction experience has been considered and consciously designed.

Can "design and the social sciences" make a real difference in making successful products and services? Maybe. To make it successful means that it must move beyond naive and traditional results of science and art. Merely repackaging their academic work and promoting it to industry is not enough. Success with industry will require that they truly understand what must be done and that they are skilled at its application. A critical element of success is the recognition that their students' success, in many ways, is more important.

What is a role for human factors and ergonomics in high technology business? The term "human factors" has several synonyms in the literature: ergonomics, human engineering, engineering psychology, operability, and usability. It is a relatively small profession in which many of the professionals are involved in the design of products (there are many human factors professional members whose fields are not product-related). The role of the human factors professional in the product design and evaluation fields is to ensure that products are intuitive to use without error, safe to operate, easy to learn, and easy to maintain.

Well-designed products which incorporate human factors principles are often said to be "user-friendly." Of course many products without human factors involvement that are not-so "user-friendly" may still get called "user-friendly" to promote sales.

Human factors works with users to determine what they are capable, and not capable, of doing, and also what they prefer to do. In the big picture of designing a sophisticated product, the number of human factors support personnel is relatively small compared to other disciplines. Product usability, however, is a very visible part of overall system performance and customer satisfaction. For these reasons, human factors should be an integral part of the design process from the early concept phase through production.

To the extent that students from the design and the social sciences become involved in human factors (and other) issues it is important that the role of software design and development be recognized as a critical skill that must be one of the "interdisciplinary connections."

There has been an enormous amount of work during the past decade in understanding "design" in the domain of software—in particular with respect to human–computer interface/interaction, but also as a generic design process such as design rationale, participatory design, situated design, and computer supported cooperative work. A significant amount of human factors work is now strictly focused on software and this and other disciplines may create a wide gap between the needs of industry and the skills of academia.

What is "Usability" and how can it impact design and the social sciences? Usability is the "Quality of use in context"—it is the effectiveness, efficiency and satisfaction with which specified users achieve specified goals in particular environments (ISO 9241–11). It is a "quality" that reflects an experience of somebody, doing something, somewhere, to accomplish some goal. Usability engineering is a relatively new discipline that provides systematic methods and tools for the complex task of designing user interactions that can be readily comprehended, quickly learned, and reliably used. Usability tests are conducted throughout the duration of a project to help select from alternative design concepts and to verify that designs meet usability requirements. Overall product and specific features are tested with representative samples of users as they use software simulations, hardware prototypes, and fully operational products. Usability testing is not the same as market research, nor is market research a substitute for usability testing.

Usability must be seen at two levels: "usability" is making sure that we are building the product right (this is what we usually mean when we talk about usability), and "usability" is making sure that we are building the right product.

The term "user interface" really refers to any attribute of a product that a user interacts with. These design attributes could include mechanical hardware, product appearance, instructional labels, user-oriented packaging, software logic, and messages. In many ways, "user interaction" better describes the process and product than "user interface."

"Usability" is one of the many commonly used, but little understood, words. Usability implies some kind of goodness of design. It is often used as synonymous with "user-friendly," "easy to use," and "easy to learn." While it is true that these are important components, they are usually thought to be "soft" and subjective. If usability is to be a concept of any value to product developers, it must be operationalized in concrete and measurable terms. Product usability is defined in ISO 9241–11 as "the degree to which specific users can achieve specified goals in a particular environment with effectiveness, efficiency, and satisfaction." Effectiveness measures the accuracy and completeness of the goals achieved; efficiency measures the resources (e.g., time, money, human effort) used to achieve the specified goals; and satisfaction measures the user's physical comfort, subjective acceptability, and attitude toward the product.

Usability can be contrasted with utility. Utility is concerned with what the product can do (features, functions, technologies). Usability is concerned with how well people are able to use that functionality—usability is the degree to which potential utility becomes actual utility. Strictly speaking, usability is not an attribute of the product itself. Usability is a relationship that exists among the product offering, the user, and the task. This dynamic aspect of usability is probably what has made it so difficult to understand, design for, and evaluate.

A usable product helps users accomplish their tasks. The actual use of the product should be consistent with the intended use. We must know who the users are, what they need to do with the product, and where they need to do it. For most communication devices, for example, there is no one profile of a typical user, nor some standard finite list of tasks performed with the device (although we know which common tasks are performed). Actually, radio and phone users are a very diverse crowd with a wide variety of communication needs.

Understanding user variance and designing for it is the key to ensuring usability. The designer needs to invest in understanding how the product is actually used by different users. A robust design will include a user-system interface that accommodates user, task, and environment variability. The appropriate focus on the design and evaluation of "usability" offers the opportunity for both designers and social scientists to learn and teach a set of rich skills that may be applied in their future execution in industry or academia.

## Human-centered design: designing the right product

Human-centered design is a product and service development process that starts with users and their needs rather than with technology. The goal is to use appropriate (adjective) technology to solve real problems, not simply to appropriate (verb) technology.

To make products that reflect a deep understanding of user/buyer needs we must truly understand that problems arise for humans in situations where they live, work, and play—in other words in relation to a background. It is in understanding these natural human backgrounds that we obtain our greatest potential for new product opportunities.

The evolution of communications technology is really a co-evolution of "communications tasks" and "communications artifacts." A task implicitly establishes requirements for the development of artifacts to support it—an artifact suggests possibilities and constraints—that often radically redefine the task for which the artifact was originally designed, and so on. To understand human activity and experience we must understand how artifacts mediate and structure action and experience. Changing artifacts change expectations—and use.

To make the right products the makers must understand (and design) "what it is for," rather than just "what it is." To understand (and shape) the phenomena that surround a new technology we must understand what it means to design. The responsibility in many cases of designing new technologies is not just that we passively observe and comment on technology—we share a much greater responsibility. Often when we create new technology, we re-create human experience. New technologies can develop within the background of a tacit understanding of human nature and human work. Technology then can lead to fundamental changes. Changes in what we do, how we do it, and ultimately in what it is to be human.

Design is the process of bridging the gap between the realm of needs and the realm of concrete expression—and this is true in all design domains. We need to be concerned not just with the "how to" of design, but with the "what" of design. Are we making the right product, rather than just making the product right?

Human-centered design has become a recognized and formal process for product design. ISO 13407, the standard for human-centered design processes for interactive systems, describes human-centered development as: "An approach to interactive system development that focuses specifically on making systems usable. It is a multi-disciplinary activity, which incorporates human factors and ergonomics knowledge and techniques. The application of human factors and ergonomics to interactive systems design enhances effectiveness and efficiency, improves human working conditions, and counteracts possible adverse effects of use on human health, safety and performance. Applying ergonomics to the design of systems involves taking account of human capabilities, skills, limitations and needs."

## Usability testing: designing the product right

Usability testing is most effective when it is done as part of a formal Usability Assurance Plan that is developed very early in the project's life. A Usability Assurance Plan is a formal document that serves to establish quantitative usability engineering requirements, a user-centered design process, and a phased user-testing plan. For many projects this approach is a radical departure from standard procedures.

Usability tests are ideally conducted to help select from alternative design concepts and to verify that designs meet usability requirements. Product usability testing is conducted throughout the product design process—from "concept" to "system optimization," through "detailed design," and "design certification" phases. Overall product usability and specific user features are tested with a sample of users operating breadboards, prototypes, software simulations, or fully operational products. Test results are analyzed to assess product usability and to identify design issues. When issues are identified, human factors personnel work with the designers from the several disciplines to develop feasible design solutions.

Laboratory-based usability testing is used to evaluate the degree to which test participants' performance, under controlled conditions, meet pre-established usability criteria. Field testing is used to better understand users' responses to, and attitudes about, the prototype or product in the context of their own natural work environment. Usability, in many respects, is a very individual thing. If a person experiences a product-system as unusable, then it is unusable. Understanding the usability of a system from the user's point of view often requires that data collection take place within the context of the user's actual work environment.

Usability testing offers an established method to improve the product—and just as importantly, it offers a method to improve the process by which products are defined and developed.

Usability testing can help to detect and resolve problems in the execution of a product concept, but it does not address the discovery of users' needs as a foundation of a product concept (Sanders 1993).

## What is design? Who is a designer?

To understand (and shape) the phenomena that surround a new technology we must understand what it means to design. Design is a very special form of human communication. For any type of high-tech (and most mass-market) products, given the real complexity of the design process, there is not one maker or designer. The process of creating may be a one-person process, however in the real world of the product, designing is always a social process.

Within an industry there are many design domains, many designers, and many who present themselves as a designer. Design is not the exclusive property of any one professional organization or group. Within most industrial organizations many engineers that work on both hardware and software, are labeled as designers, and operate in industrial design, mechanical engineering design, electrical engineering design, software/human computer interaction design, etc. This applies also to non-product design fields, such as architecture, urban design and systems.

Furthermore, I would argue that designing is not just a visual process—rather, in important respects, it is a verbal process. This does not mean that visuals, models, and software (as end products) are not the reason for design—they are—but that the process, when properly executed, is inherently verbal.

## A field of dreams

Many companies have recognized that the "field of dreams" approach to product development will no longer work: "If you build it, they will buy it." They may not yet have recognized the logical fallacy: "If they buy it, then you built the right product." Too often product companies confuse market success with design success (*post hoc, ergo proper hoc*). Now, successful sales do not mean that you built the right product, right. It may just mean that (some) customers need (something like) it now and want your company to sell it now, and it may just mean that a change in competition or user needs tomorrow will change what they want and how they buy it. We have an opportunity to redefine this logic to "If you build the right product, and you build it right, then they will buy it."

## Designing is not just physical

In real-world product and service design and development, there are probably very few real-world scientific constraints that one encounters. Constraints are usually human constructs – "it can't be done"; "over-night physics" solve it; it is stopped, or started, by "engineering-ego," etc. Previous failure is a common source of design constraints. These archeological artifacts, usually based on the misapplication, misinterpretation, or unnatural evolution of poor design or failure, can become the dynamics for rejection.

The interpretation of design goodness, and its impact on future design can range from proof, to tacit knowledge, to the use of bargaining chips, and beyond. Likewise, the "facts" about what a customer wants (probably doesn't really want), can be based on company opinion, intuition, casual personal experiences, and folklore.

Designing is not just physical. Truly successful design must be driven by the successful scientific contribution which resolves (at some level and time) the social, organizational, personality, anthropological, etc. needs of both designers—and the ultimate users. The potential for future, successful, social science industrial improvement (not just its mere study!) is major. Clearly, the potential for real fun in both the futures of academia and industry is there.

## Success for academia and industry

The present chapter has focused on just one part, of one possible direction, for one possible industrial domain. The challenge is to alert and entice faculty, and to fascinate students to this domain as a real, dynamic, and incredibly appropriate interdisciplinary target.

The interdisciplinary integration of anthropology, psychology, sociology, with art and design is not the same as the interdisciplinary integration of anthropology, psychology, sociology, and art and design. The impression of an interdisciplinary funnel of social scientists which feeds the design needs of artists is radically different from the impression of the dynamic collaboration from a potential transdisciplinary team. Academic success with industry will require academic success with academia. The recognition that mere art-and-craft-based structural models is inadequate for education and, presumably, for successful future industry employment, also should reflect the inadequacy of much traditional social science training for successful future industry employment.

Yes, a communications artifact may be a product of the artistic, but this object of design is much more than that. The incredible potential future for academia and industry (and, of course, for customers) requires, in my judgment, that the reality of the process of design as a social process replaces the tradition of creating as a one-person artistic process.

Finally, the hope is that the Bachelor of Design program is not just an art and design program that is guided and trained by the social scientists, but that appropriate social science students are trained and graduate with that program. In addition, while the present realistic target is at the undergraduate Bachelor level, industry can only hope that graduate-level programs and graduate-level students are integrated in the not-too-distant future.

We (academia and industry) need to break our focus on the product-as-object. We need to understand and design what it is for, rather than just what it is. To understand, and shape, the phenomena that surround a new technology we must understand what it means to design. Our responsibility here is not as mere passive observers of, and commentators on, technology—we share a much greater responsibility—we create new technology, and in doing so, we re-create future human experience.

## References

Crawford, C. M. (1994) *New Products Management* (Boston, Mass.: Richard D. Irwin).

Sanders, E. B.-N. (1993) "Product development research for the 1990s." *Proceedings of the Human Factors and Ergonomics Society, 37th Annual Meeting* (Santa Monica: Human Factors and Ergonomics Society), 422–6.

# 4 People-centered design

## Complexities and uncertainties[1]

*Jorge Frascara*

### Introduction

The proposed link between design and the social sciences is the result of two things: the recognition of changes in the design profession in connection with both purposes and methods of operation, and the perception of the way in which these changes require an increased participation of social scientists in the conception and development of consumer products and public services. To elaborate on details, I will discuss a series of headings and brief explanations, regarding those dimensions of design that lead to new understandings of its nature and its practice.

### Objects and people

While design has been traditionally concerned with objects and processes, we have to recognize the impact that those objects have on people. We have to stop thinking of design as the construction of graphics, products, services, systems and environments, and think about those as means for people to act, to realize their wishes and satisfy their needs. It is the needs and the wishes that we have to serve; the objects of design must be seen only as means. This requires a better understanding of people, of society, and of the ecosystem, and calls for an interdisciplinary practice.

### Operational impact and cultural impact

Every design project has an operational objective: it is supposed to affect the knowledge, the attitudes or the behavior of people in a given and desired way. But any object deployed in the public space, be it communicational or physical, has a cultural impact. This cultural impact affects the way people operate with people and with things, and contributes to the creation of cultural consensus. More has to be done to understand this cultural impact so that designers can operate more responsibly in society.

## Communication as partnership and negotiation

If visual communication design is concerned with affecting the knowledge, the attitudes and the behavior of people, then it should do this in an ethical way, that is, seeking partnership in the process of change, rather than communicating things to people. The process of communication should be seen as a process of negotiation where the position of the originator of the information and that of the interpreter enter in contact searching for a common terrain. Unidirectional communication is unethical and inefficient, and it promotes a passivity that in the long run will weaken our civilization. In ethical communications, one communicates with someone about something; one does not communicate something to someone. In ethical communications the popular terminology, borrowed from electronics and information science, that defines the poles of the communication chain as transmitter and receiver, is untenable. In a universe of people engaged in communicational exchanges, it is more fitting to talk about producers and interpreters than about transmitters and receivers, terms that do not allow room for context, history, expectations, goals, values, priorities, feelings, preferences, cultural differences and differences of intelligence. In ethical communications, the producer has to speak a language that the audience can understand. If producers really want to communicate, that is, if they want to be understood and not just be listened to, they should remember that people can only understand things that relate to things that they already understand, and that it is impossible to communicate, therefore, without using the language and the experience of the audience. This is why the ideal form of human communication is the dialogue, where the interaction allows for exchange and adjustment, and for the building and extending of a shared terrain. Given that in our society a designer fashions communications for mass media, attempting to reach thousands of unknown people, it becomes clear that design needs to liaise with the social sciences to improve our understanding of the audiences and increase the possibilities for both ethical and effective communications and products to come to exist.

## Accountability

There is a need to develop a more accountable design practice, where the judgement of quality should not depend on the coincidence of several subjectivities but where it could be based on actual measurable benefits: human, financial or other, that could be expressed as the return of the design investment. Without attempting to reduce all design activity to only those dimensions that could be measured, it is time to be serious about design as investment, so that it is not seen as an expenditure, and a superficial one.

## Public good

Much has been said about doing voluntary work for public service. This is not my point. The public good must be the most important objective of design activity, and it should be sought with the best resources, being understood as an investment with high returns affecting hidden dimensions of the economy. Let us just think about tax return forms, which are filed by some 20 million people in Canada, and about the Australian Communication Research Institute estimate that each error made by a form-filler costs about $14 to the administrating agent. Think about the CDN$ 300,000 that a spinal cord injury costs in the first year of treatment to the Canadian health care system, and the number of car crashes that result in spinal cord injury. Think about the 200,000 Canadians that suffer injuries every year in traffic collisions and the 51 million days of work lost every year in Canada due to injuries in general. Much can be done about this with well-designed communication campaigns, information improvement, public education and concerted programs, but to do this we need the best brains in a variety of fields, centering, of course, on design and the social sciences. And we also need to make substantial investments in these programs. There is experience in this terrain: the traffic safety campaign produced in the state of Victoria, Australia in 1990, required A$ 6 million dollars in media, but the government agency that acts as third party liability insurer saved A$118 million in compensation payments during the first year of implementation of the campaign. This happened nine years ago, and by 1999 the combination of broad scope police programs and communication campaigns has reduced the injury collision levels to that of the 1950s, about 50 per cent of what it was in 1989.

## Relevance

Design has to be relevant so as to raise above fads and fashions and penetrate all dimensions of life with a view to improving it. Irrelevant design is a liability for the profession and the environment. If we are looking at strengthening the position of design among other human activities, we will have to review the relevance of design projects and foster work in those areas where design could actually make a difference for the better.

## The designer as a problem-solver versus the designer as a problem identifier

Design is a problem-oriented, interdisciplinary activity. There is a need to identify important problems and develop interdisciplinary strategies to deal with them. It is not sustainable to continue just reacting to clients' requests for design interventions. It is necessary to consider the discovery and definition of physical and cultural problems as an essential part of design. The nature of each problem might suggest the spectrum of disciplines required to confront it. A set of tools to look at the world will have to be developed by inquisitive, critical, interdisciplinary observation, performed by people in love with humanity.

## Sustainability

Given the state of affairs today, in both environmental and cultural terms, it is not possible to design anything without framing it within the notion of sustainability. The escalation of waste and the generally irresponsible attitude of industry and governments about the use of toxic and damaging products and processes is as bad to the physical environment as the promotion of violence and selfishness by the entertainment industry is to the cultural environment. Cultural and physical sustainability must become part of every design process, and schools will have an important role to play in the formation of the new generations of designers. While it is easy to recognize blatant inequalities at an international level, there is a lot to do concerning cultural sustainability in our own backyard.

It is not sustainable to continue just reacting to clients' requests for design interventions. It is necessary to consider the discovery and definition of physical and cultural problems as an essential part of design. The nature of each problem might suggest the spectrum of disciplines required to understand it. A set of tools to look at the world will have to be developed by inquisitive, critical, interdisciplinary observation, performed by people with an ability to understand large contexts and long-term goals.

## Efficiency and democracy

These are the two poles between which collective decisions take place. Where to strike the balance, is a matter of judgement. If design decisions, and other decisions that affect us all, are to be collective decisions, then more should be discussed about their nature, and about criteria for appropriate balances between the two poles. To get things done there is no alternative, and many times excellent design ideas do not get realized because the originators lack the necessary ability to interact with people in decision-making positions. The more original the design proposal is, the more difficult it will be to implement it, particularly when needing significant investments. Designers at the forefront of the profession are understanding design as the design of interactions between people and objects; now we have to develop a better understanding of the interactions between people and people.

## Planning and self-organization

Every design problem involves these two aspects: it is not possible to continue believing that planning can solve it all, or that self-organization can solve it all. Understanding the capacity of things and people to organize themselves when thrown in a given situation puts planning in an interesting perspective. The sport team model or the Panzer Division are not sufficient, and the Calcutta model is not desirable or sustainable. In design we should look more at how things interact with each other, and at the dynamics of large complex systems, such as cities, ecologies or the stock market, and look for better conceptual models to replace old planning strategies.

Design is a problem oriented, interdisciplinary activity. There is a need to identify important problems and develop interdisciplinary strategies to deal with them.

## The complex and the complicated

I like to make a working distinction between these two terms, for the sake of a clearer use of them. Let us say that a computer network is complicated, since it consists of a great number of parts but it is possible to account for them all. Social relations, instead, are complex. Aspects interact with aspects and constantly change, preventing us from ever developing a perfect description, definition or explanation. Our relations with the complex are always in a state of flux. As soon as we have people in the equation, the problems become complex. Design, more often than not, is a complex problem.

## Form, materials and self-expression vs. context and content

The major preoccupations of the avant-garde, fine-arts-based design education of the 1920s were form, materials and self-expression. Materials exploration was added by design education to the self-expression/form exploration approach of the avant-garde fine artists. We now realize that graphic design education today, some 70 years later, is in the main concerned with the same issues. This goes against the development of an awareness of content and context, and of the way in which these two dimensions should condition design action. We have now enough collective experience on form and materials; we need to transform self-expression into resourcefulness and inventiveness regarding the visual language in order to be able to speak the language of the public being addressed. We need as well to concentrate on formalizing and codifying the problems of content and context, learning and teaching how to transform them into elements of the design process.

## Objects of thought and dimensions of judgement

Public opinions and attitudes are formed on the basis of these two dimensions, that for the purpose of persuasive communications must be distinguished by the designer. Objects of thought are things we have in mind, and these could be concrete objects, such as the Eiffel Tower, my mother, or the concept of leisure. Dimensions of judgement are value axes: good–bad, useful–useless, dangerous–safe, attractive–repulsive, etc. The positions of objects of thought on specific points in scales of dimensions of judgement forms the basis of attitude formation and attitude change.

## The dematerialization of design

All the above emphasizes the fact that as designers we have given up our exclusive obsession with products, materials and manufacturing processes, and have become more concerned with the contexts in which objects and communications are used by people, and with the consequences that the existence of those design creations have on people in general. We have moved, for instance, from the design of work stations to the design of work. It is not possible to invent and design the perfect chair on which a person could be sitting eight hours a day, five days a week, without becoming physically fatigued or handicapped in one way or another. It is wiser to design a work pattern which, including the design of furniture, would be centered on the design of the activities to be performed. We have moved from the design of teaching aids to the design of teaching situations. The success of a learning experience cannot be trusted to the design of a teaching aid. The activity has to be planned so that the teaching aid contributes its best to the experience. Many factors affect this terrain, but certainly the teacher's actions, the student's actions and the environment in which the intervention occurs, all contribute to the learning event and must be seen as part of the design problem. All this points at areas were the association between design and the social sciences promises to equip us to be able to face a number of complex design problems with better sets of tools.

## A word of warning

Design, particularly in its relation to human factors and other aspects of the social sciences, has been connected both in the recent past and in my own words, to the notion of efficiency. Efficiency is central to design discourse. Why to design, otherwise, if not to increase the efficiency of something? To facilitate the satisfaction of needs, to multiply force, to increase comfort, to extend our abilities, design is conceived as the way to bring efficiency to our lives. Efficiency, however, has been excessively emphasized in our culture, and particularly in North America, and our lives are being destroyed by processes of downsizing administrative supports, increasing performance expectations, and turning into businesses what used to be public services. All this puts pressure on people to be efficient, to work harder and to feel guilty when taking a stroll on Saturday instead of dealing with the workload one takes home every weekend. It is imperative to look at the contexts within which design operates, and at the value systems design promotes, and see in design also the possibility to respond to a broader set of human needs.

I recognize three areas for the practice of design: design that works to make life possible, design that works to make life easier, and design that works to make life better. I work very much on the first one. My work on traffic safety communications intends to keep more people alive, and to reduce pain and suffering. Like the medical profession, I concentrate on physical survival, the biological in us.

Design that makes life easier is the design of tools that multiply our force, clothes and other systems for climate adaptation that keep us at near to ideal temperature, and any object that extends the abilities of our body.

Design that works to make life better, includes sensual and intellectual enjoyment, the promotion of mature feelings, ability to reach high degrees of consciousness about our lives and our actions, and cultural sensitivity to build civilization and relate constructively to others; all those things that make us specifically human. These three dimensions, of course, can many times merge in the same project.

I see the relation between design and the social sciences as one that will increase the ability of design to deliver efficiency. But I see the need to use that efficiency so that we can increase not only our production of consumer goods, but our time to reflect about our human condition, our time to interact with others we love, and our time to introduce more significance, enjoyment and consciousness in our daily actions.

I hope this proposal of connecting design and the social sciences will promote the possibilities that the formalization of this interdisciplinary connection creates, and that it will also promote a critical look at the contexts within which we operate, ensuring that a humane life is held at the center as the maximum aspiration of any intellectual effort.

## Note

1 Some aspects of this article were published in "The dematerialization of design/La desmaterialización del diseño," *Tipográfica*, 50, 18–25, Buenos Aires, November 2001.

# 5 Social sciences and design innovation

*Bernd Meurer*

## Introduction

Design must take three major factors into account: first, that the individual scope of action has been significantly enlarged and will grow in future; second, that knowledge plays an increasingly important role in the economy; and third, that the globalization in the economy goes along with the individualization of society. The fact that knowledge has become the most important productive force, and that capital today as human capital, that means as fantasy and eagerness to learn, is more important than material capital, shows how far the basis for "the individual as protagonist, as designer, as juggler and director or stager of his biography, his identity, his social networks, his relationships, his convictions" (Beck 1993) has matured. This also shows in the urban design of many middle-sized cities.

The return of the public to the city in interrelation with individualization does not necessarily mean de-socialization or social isolation. Individualization comprises "first the dissolution and second the replacement of industrial forms of living by other forms, where individuals have to produce, to stage, to stitch together their biographies themselves." The release of man from traditional ties makes it necessary to test or try new forms of social behavior. "Thinking of oneself and living for the other, although seemingly contradictory, turns out to be a meaningful inner correlation. The one who lives for himself, must live socially" (Beck 1997).

The private is not anymore a stronghold against the public. The place of living has become a stage for the public and the public has become a stage for the private. We do not know how individualization and socialization will unfold. We only know that it is important to design processes and artifacts that allow us to generate these correlations. The question is: how can the desire for self-determination be brought into accord with the desire for the split common ground, which is just as important? This confronts us at two levels that are interwoven: namely at the social level regarding living and at the material level regarding objects.

The Museum of Modern Art in New York has recently organized an exhibition about new forms of living; it was titled "The Un-Private House." Below, I refer to three examples from this exhibition.

## The private and the public

If we talk about the end of unambiguity (Zygmund Baumann), about distraditionalization (Anthony Giddens), about the flexible man (Richard Sennet), about the relief of work by other activities (Jeremy Rifkin), about globalization that comes along with individualization (Ulrich Beck); and when we talk about how these social changes take place, in what relation to the transformation of products and processes, then most happen in the non-spectacular world of everyday life—a process of impenetrable ambivalence.

I will concentrate in this chapter on the transformation of the public and the private in architecture and design. At the end I will give three examples of design as related to such transformation—three examples of how far the radical shifts in the public and the private have already entered the design of housing. Of course, the examples appear still as exceptions. On the other hand, they are not truly exceptional anymore. The design debate is already penetrated by these changes, even if partially in the subconscious.

New and unusual forms of living develop in a way that is different from those examples of urban transformation that of their own, for instance because of technical innovation, create new forms of use and behavior. While living is changed by new technologies, it depends more than other fields on cultural traditions. It is influenced by habits that have developed over thousands of years and did not change substantially (such as eating, sleeping, etc.). On the one hand, living is shaped by conservative habits, to which man holds, especially in times of radical change and upheaval. On the other hand, in its most developed form, living is characterized—also visually—by the dissolution of the sociological base of the society and the outcome of new ones.

The world of objects constitutes itself in activity. We perceive objects by nothing other than use, in the most far-reaching sense. Use is activity. In creative use, say design, we are confronted with two different kinds of transformation. On the one hand, there is conscious design, that appears in innovations—for example in social intervention, in new buildings, in strategies for mobility, etc. On the other hand, there is subconscious change, imperceptible in everyday life where considered change is mingled with not particularly thought about activities. In a sense, the transformation of our artificial environment is a systemic chaos, where all activities are related and in continuos flow—overlapped by planned strategies and innovations.

Materially, the artificial environment appears as something different. It appears as a gigantic accumulation of artifacts, as an arsenal of buildings, streets, spaces and all the other material products, which are part of life and of the economy.

In the traditional understanding of architecture and design, the transformation of life is seen above all as material change. The change of the material world comes to the fore. Seen in this one-dimensional way, the development of civilization is reduced to the material. Our world is object-fixated. And this object fixated understanding of architecture and design is also about allowing space for social and technical processes. But consciousness of human and social activity is thereby narrowed to the category of the mere usability of artifacts.

This division between object as a static thing and process as a dynamic action has its roots in pre-industrial design methods. In the pre-industrial design history objects don't have a self-dynamic momentum of their own—with the possible exception of, say, wind and water mills. This static understanding of artifacts still continues to have a strong effect on design—especially in architecture, where the work of an architect is thought more as an unchangeable object than as a transforming process, even if space defines itself in terms of time.

In the course of mechanization and industrialization, the concept of dynamic action processes slowly found its way into design, but mainly in a purely technical way. Mechanical engineering, transportation and communication are examples of areas where this first developed. However, generally the task of design is still not dedicated to processes but rather to the object. The immaterial—the processual—of it is merely considered as a service function. With the microprocessualization of everyday life, the knowledge orientation of the economy and the growing ecological awareness, the processual is pushed to the fore. This appears at the base of social behavior. The passive and static industrial biography dissolves, and the active do-it-yourself biography grows. With this development the dynamic action processes place an ever-increasing importance on social, economic and design innovation.

In research and development of mobility, for instance, thinking in terms of object-process-systems has already become an everyday method. After all, mobility research and development as a whole includes a lot of processes and products. There are product innovations (like the development of new vehicles and accompanying objects), infrastructural measures (in the field of highways, tracks and communication lines), the development of processes (like intensifying product-use by, say, car-renting and car-pooling), the development of new organizational structures (between transport services, political authorities and the users) and the development of new services. The decisive difference with ordinary product (and process) development is that various tasks are no longer handled separately from each other but rather in an interrelated object-process-system fashion. But still most of the design results are thought of as unchangeable things that don't move or transform. The results reflect more the fundamentalist culture of certainty than the art of doubt.

The invention of comprehensive object-process-systems has become a key for the social and economic modernization of our life-world. Anyone who has a nose for this networking innovation will not only be successful in economic, social and ecological innovation but will also—putting it in broker oriented terms—earn high profits in the stock market.

The transformation of the working world changes the social and spatial organization of work and with this the social and spatial organization of our environment. The development of design and architecture must take three major factors into account: first, that the individual scope of action has been significantly enlarged and will grow in the future; second, that knowledge plays an increasingly important role in the economy; and third, that the globalization of the economy goes along with the individualization of society.

The fact that knowledge has become the most important productive force, and that capital today as human capital (Gorz 1998)—that is, as fantasy and eagerness to learn—is more important than material capital, shows how far the basis for "the individual as protagonist, as designer, as juggler and director or stager of his biography, his identity, his social networks, his relationships, his convictions" (Beck 1993: 151) has matured. All this has far-reaching consequences for the comprehension of living and working, and of the public and the private. In the course of the individualization of work and working processes also the working environments individualize—both despite of and because of their networking. Additionally the social splitting of job, housework, family work and social work, gives way increasingly to a plural occupation of the individual. If one looks at the problem of housing, the Dutch architects group MVRDV argues: "The demand for greater variety and even more extreme forms of dwelling is gaining momentum. The ideal home no longer exists: there are thousands of ideal homes" (MVRDV 1997: 4). The changes in practice and concept of living, working, the public and the private, require the dissolution and alteration of the (industrial) typology of space, of buildings and of city-zones and the other things of life.

The modern, knowledge-oriented economy is based on the development of lively, cosmopolitan and urban milieus to gain higher synergy effects. This economy-oriented interest corresponds with the social oriented idea of a complex urban structure where the most different fields of activities are tied in an attractive way.

Also in the choice of living surroundings grows a renewed desire for urbanity, which in the past had been driven out of people's mind by the noise and emission problems. In the more recent past downtown housing has become more and more attractive. That is—apart from the search for new types of living—urbanity in a new comprehensive sense is growing because the product, namely the inner city, has changed. Downtown areas in Western Europe—and in a different way also in America – have gained an ambivalent social attraction with a growing number of sidewalk cafes, restaurants, attractive spaces, modern cultural set-ups and changing closing times.

Unlike a few years ago, public space is no longer a commodity the loss of which would be lamentable. Understood as a dense meeting space full of action, it slowly becomes a renewed subject of planning. Interestingly enough, the renaissance of the public space happens at a time when social communication—if we see it purely technically—no longer needs to rely on public urban space. Today, one can communicate with people and communication machines all the way around the globe without ever leaving one's home. Nevertheless the need for real communication in real public spaces is growing in addition to virtual communication and spaces. Even increasing portability of communication means (like cell-phone, notebook, etc.) does not damage this renaissance. The contrary happens: as of recently, telecommunication is no longer limited to stationary communication equipment tied to the private world of working and living. In the public space the individual is today confronted by the behavior of others not only in a spatial context but also through telecommunication elements.

At the same time the microprocessualization of everyday life has shifted the division between the public and the private. For years past, public space became global and penetrated private space. With television, public space became omnipresent. Private space in the shape of the private car has occupied street space and driven out public activities such as children's play and leisure. The public street space mutated to a gigantic accumulation of four-seated private spaces on wheels. Today, pedestrian areas are slowly growing from small islands to larger car-free city areas. The spheres of the private and the public are penetrating each other. This shows in the development of many mid-size European cities, such as Freiburg, Strasbourg, Tübingen, Bologna, and Montpellier, to mention just a few. For instance, the redevelopment of Montpellier to a so called "technopolis" included the transformation of the whole downtown, namely the entire old part, to a pedestrian area with thousands of open-air restaurant tables, beautiful shops and a lively city center. This process is in itself ambivalent. It includes social hardships. The city also changes towards a place solely for consumerism and services whereby the traditional population may be driven out.

But the return of the public to the city takes place in interrelation with the individualization of the society. The consequence of individualization is not necessarily de-socialization or social isolation. Individualization implies "first the dissolution and second the replacement of industrial forms of living by other forms, where individuals have to produce, to stage, to stitch together their biographies themselves" (Beck 1993: 150). Individualization involves self-authorization and self-responsibility. The growing social demand for self-determination and self-consciousness comes out of an increased freedom that questions the traditional forms of living and demands to try new principles of common ground. More than ever the individual is forced to make his own right choice on social behavior. Naturally this also increases the risk of error. In the end modernization is a revisable, and not an un-revisable process. The dialectics of modernization and counter-modernization have to be permanently dealt with.

The release of man from traditional ties makes it necessary to test or try new forms of social behavior. A new culture of common ground is growing; for example, when citizens take over tasks of the state like nursing care and senior citizens care and—on a different scale—the work on the "local agendas." In the world of economics such a change is also noticeable. Some businesses understand that they are more than money partners for their employees. While the social benefit state becomes less affordable, a growing public spirit emerges. Little plants, certainly; but they grow even in the private economy. We experience at the same time both degeneration and a regeneration of a responsibility in ethics. The sociologist Ulrich Beck noted that "thinking of oneself and living for the other, although seemingly contradictory, turns out to be a meaningful inner correlation. The one who lives for himself, must live socially" (Beck 1997: 19). The newly developed spaces for social interaction in the cities are not only areas of consumption but also test grounds for these social experiments.

The private is no longer a stronghold against the public. The place of living has become a stage for the public and the public has become a stage for the private. We don't know how and in what form individualization and socialization will unfold. We only know that it is important to develop processes and artifacts that allow us to generate this correlation. Ulrich Beck asks: "How can the desire for self-determination be brought into accord with the desire for the split common ground, which is just as important? How can one be at once individualistic and absorbed with the group" (Beck 1996: 12). This question confronts us at two levels that are interwoven: namely at the social level in respect of living and at the material-processual level in respect of objects. But the industrial past thoroughly shaped the terms of our work, life, consumption and leisure style, partly by the hardening of obsolete thought patterns. The difficulty is to replace the old industry-oriented functions with the development of new knowledge-oriented forms.

Last summer (1999) the Museum of Modern Art in New York opened an exhibition on new forms of housing with the intention to indicate how the social changes result in new forms of living. The exhibition was also to be seen until April this year at the Museum für Angewandte Kunst in Vienna. The subject was provocatively titled The Un-Private House. In this title the un-private does not stand for the public. It rather means the broadening of the private into the public, the change of the private that gives space to a new type of the public in the private. The public enters the private and the private enters the public. A changing sphere where the private dominates the public and the public dominates the private. Many of the examples shown still glue to the living form of the detached house. The rethinking of the detached house has not yet quite developed. Unquestionably the wasteful land usage for one-family housing in suburbia is still much more dominant than the yearning for urban living forms. Similar to the hang-ups of business in the mono-functional business area, it is a question of understanding as well as promotion on the part of municipal authorities, for change to come about. It is a question of what kinds of synergy-producing product-process-systems are more useful and attractive.

The one- or two-family house is expanding with its consequence of landscape destruction. By official predictions the world will in 50 years be a total mass of houses. What today seems to be self-evident, the spatial splitting of the city in living, working and leisure, started only 150 years ago. Prior to that, work and living was combined. With the industrialization of work, the work emigrated from the place of living as labor into the workshops, into the factory, into the office. The place of living was transformed into a place that is solely for living, devoid of any work aspects. In connection with this began the bourgeois retreat into the private. But this form of living which emerged only about 150 to 200 years ago is hardly to be considered as historically the last and final.

Many of the examples this exhibition shows already test the un-private, but still related to the antiquated structure of the detached house. They show the penetration and the overlap of functions that dissolve the traditional floor plan. I want to refer to three examples. First, a house on a spacious lot similar to a park; second, a house in a block on a very small lot, that has—in itself encapsulated—four and a half floors; and third, a house where the lot is almost completely taken up by the house in a high-density area.

The example from the Dutch architect Ben van Berkel works according to the principle of the Möbius loop. The house is conceived for a couple with children (Figures 5.1–5.3). The two adult inhabitants work at home and want to also have the opportunity to withdraw from each other. The designer tries to make living, working and sleeping meet in a new way. An endless loop or band forms two intertwined sequences of spaces. Semi-public spaces, two different work spaces, family spaces, and private spaces, are in an open relation. The circulation areas not only serve as a function but also acquire a temporal dimension, reflecting integration as well as separation of living, working and sleeping. Instead of creating traditional spaces and forms—following a static idea of functionality—the house articulates the changing relations of the inhabitants. The public and the private are in a continuos flow.

The spatial characteristics of the Möbius loop are visible in the plan and the section of the house. The loop-like breakaway from average geometry

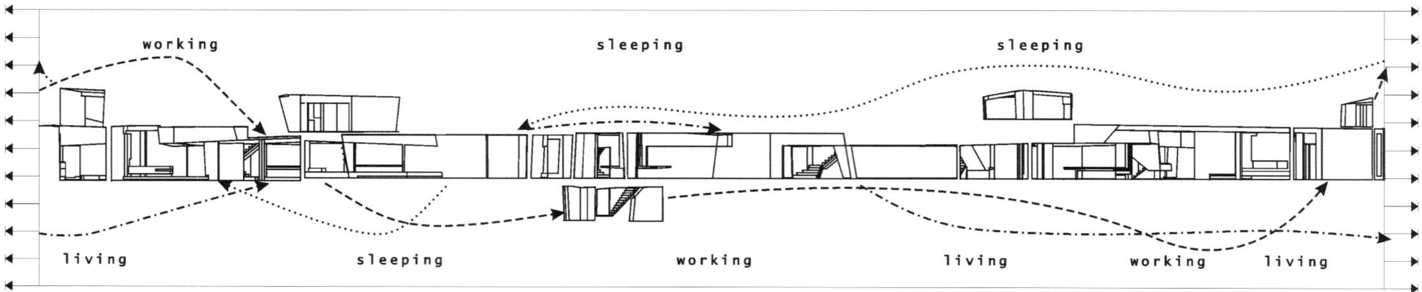


*Figure 5.1* Ben van Berkel, UN Studio van Berkel en Bos, Möbius House, construction sketch (reproduced by permission from UN Studio van Berkel en Bos).

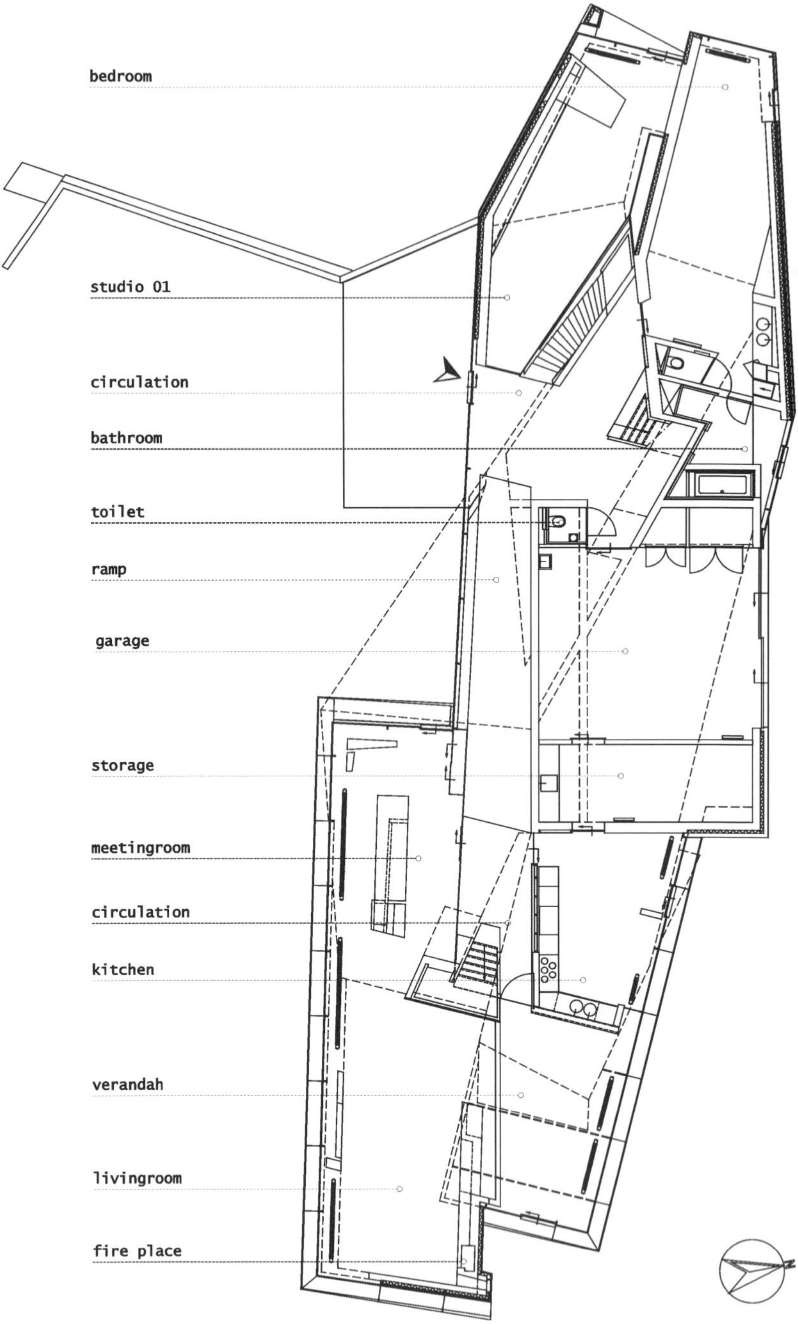


*Figure 5.2* Ben van Berkel, UN Studio van Berkel en Bos, Möbius House, section (reproduced by permission from UN Studio van Berkel en Bos).

*Figure 5.3* Ben van Berkel, UN Studio van Berkel en Bos, Möbius House, photo Christian Richters (reproduced by permission from UN Studio van Berkel en Bos and Christian Richters).

brings about space to respond in a new manner to the 24-hour rhythm of living, working and sleeping. The occupants can go for an endless Möbius-trip in this house. The twisted band turns the inside to the outside and vice versa. The intertwined spaces and structures seem to fold back the design over onto itself. The house articulates the complex relationship between the occupants themselves, in their relation to the house and to the impressive nature of the site.

The house is conceived as a detached house. In this respect it represents an antiquated form, despite its novel features. The detached house has become a nature-destructive mode of living. On the other hand, in this example, the house conserves a suburban enclave outside the city. A new form of space arrangement collides with the land-eating use of the city. As a detached house it is far from creating an alternative to the wasteful use of nature in housing. In the contradiction of both aspects, namely the modernization of the space arrangement by using the Möbius band in a new way, and the preservation of the antiquated model of the detached house, appears the inconclusive dialectics of modernization and counter-modernization. The modern societies are confronted by the principles and the limits of their own models.

Design has to show effect. And this effect must be strong enough to be felt. Seen in this light the house of van Berkel opens itself to the ambiguity and the ambivalence of its design. Here design has some points of contact with the skepticism of modern sociology. Not the dissolution or neutralization of conflicts should be the aim, rather their civilization.

As a second example I want to refer to a house by the Dutch group of architects MVRDV (Figures 5.4 and 5.5). This example deals with high density house-building in a block. The architects tried to work on the design problem of planning a house for living on a small lot and the sociological problem of the individualization of society. "The society moves closer together," writes MVRDV, "the private mercilessly is combined with the public, the marital row is fought out in public and the intimate conversation takes place on the street by cell phone. Only the neighbors you don't know, since soundproofing became much better" (MVRDV 1998: 43). The old borders between the public and the private have become open. New ways of living together and living apart seem to be suggested in these new forms. We have to invent other distinctive features. Individualization is an intermediate stage in this process. From the historical base of the passive industrial society emerges the reflexive, the active knowledge-oriented society.

For the designers, the sociological aspect does not explain the architectural quality, it serves as a means of understanding the spatial structure of the house—particularly as in this example—where it was achieved by using traditional methods in a new manner. The lot is very narrow, with adjoining multi-storey houses on both sides. The design intends the greatest possible spaciousness and versatility within the limited total space. The house is only 4.2 meters wide and 16 meters long. Within the height of 9.5 meters normally only three floors are possible. This design achieves four floors while the space over the floor in much of the building is two or three storeys high. To compensate for the reduced dimensions, the house is conceived as one long space that develops over four and a half floors, offering closed privacy as well as open un-privacy. There are solids and voids encapsulated in one another. The un-private alternates with the private. The bedroom – the room with the highest degree of privacy—is located between the working level on the fourth floor and the spatially connected living space on the second floor. This again is spatially connected with the dining area, at one and a half storeys lower. The stairs are conceived in a self-contained staircase which develops in an irregular manner and creates an artistic form in the living area.

Open sections of two and three storeys alternate with sections of narrow ceiling height. The dining, living and studio spaces seem to be fluid. They appear as broken or blind continuations of themselves. They develop from the two-and-a-half-storey veranda over the balcony for the living area one storey higher up to the terrace above the bedroom. All this is done in a bewildering or surprising manner so one never sees the whole extension of voids and solids. The ambivalence of this effect takes place apart from the pure geometric clarity of space-structures. A series of non-enclosed rooms has been created with varying height and privacy, each space combined in an individual relationship with the exterior.

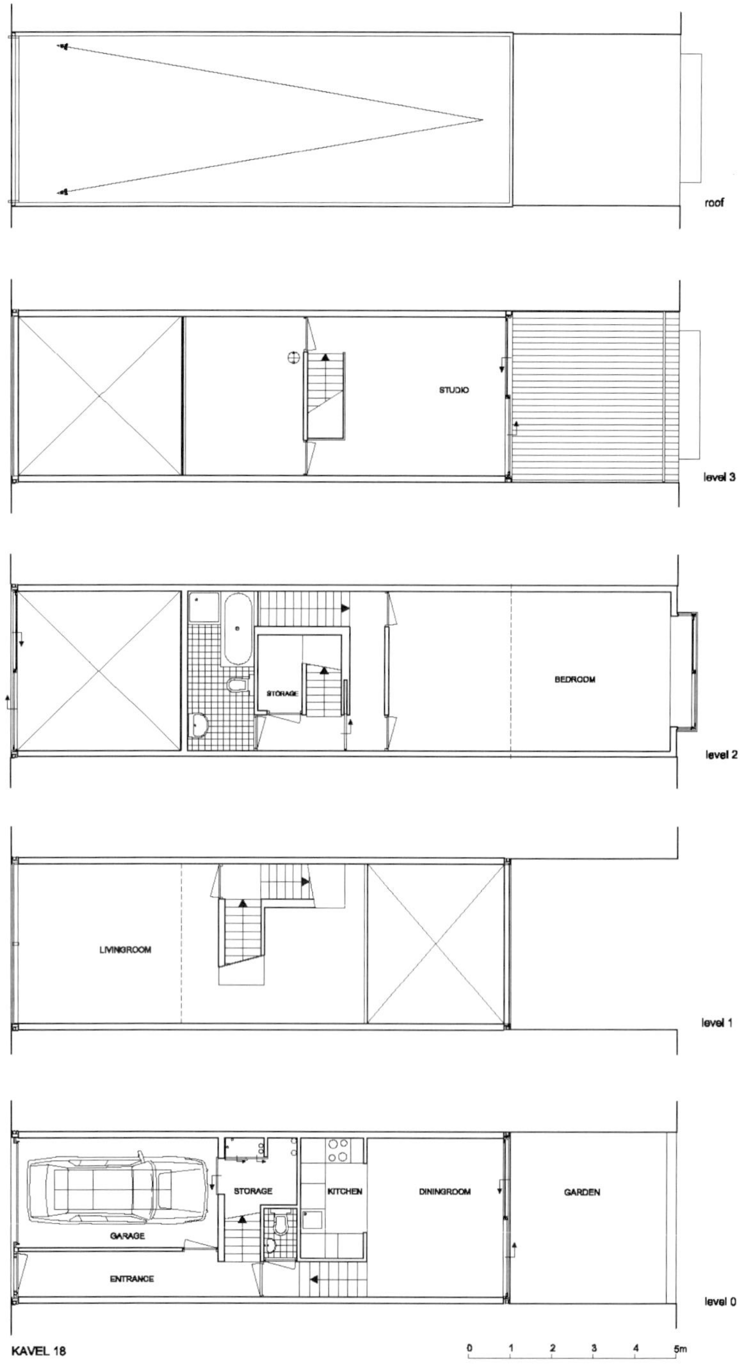


*Figure 5.4* MVRDV, Borneo House Number 18, floor plan (reproduced by permission from MVRDV Architects).

*Figure 5.5* MVRDV, Borneo House Number 18, photo Nicholas Kane (reproduced by permission from MVRDV Architects and Nicholas Kane).

*Figure 5.6* Shigeru Ban, Shigeru Ban Architects, Curtain Wall House, photo Hiroyuki Hirai (reproduced by permission from Shigeru Ban Architects and Hiroyuki Hirai).

As a third example I choose the curtain house of the Japanese architect Shigeru Ban (Figures 5.6–5.8). In plan it seems rather traditional. But the emphasis of the innovation in changing the function of the façade radically transforms the use of the house. Positioned in a dense residential neighborhood in Tokyo it is designed as a three-storey house. The entrance and the working and parking areas are on the ground floor. Living is on the second floor and sleeping on the third.

*Figure 5.7* Shigeru Ban, Shigeru Ban Architects, Curtain Wall House, photo Hiroyuki Hirai (reproduced by permission from Shigeru Ban Architects and Hiroyuki Hirai).

It involves the transformation of traditional life-habits. Half of the façade consists of a two-storey outdoor curtain that is combined with sliding glass partitions that form a terrace about 2 meters reset from the curtain. Without a doubt this house generates space for the most open and at the same time the most secluded life forms. Over and above the aesthetic and material qualities of a modern house the space becomes public stage,

*Figure 5.8* Shigeru Ban, Shigeru Ban Architects, Curtain Wall House, photo Hiroyuki Hirai (reproduced by permission from Shigeru Ban Architects and Hiroyuki Hirai).

private refuge or something in-between at the option of the occupants. They can change the character of the house by making one of four main choices: first, a totally open option, where the private is in the public; second, one where the inner is semi-closed from the outer by transparent sliding partitions; third, one where the visual contact with the public is interrupted but the acoustical is not—by closing the two-storey-high curtain; and fourth, one where the partitions are closed in addition to the curtain so that a hermetic, divided enclosure is created.

Closing only the curtain combines the living with the sleeping area by creating a continuous inner space between the two. Also one has the impression of living in a maisonette. The house has the possibility of making bedrooms, kitchen and living room into public rooms, where they have no division from the public other than by (stage) elevation. To make the rooms private you close the transparent partitions. They also isolate against cold, heat and noise. If you close the curtain you are protected visually. The degree of the private character of the interior spaces and their public exposure—or the participation in the public—are not unlike those in a traditional Japanese house. The functions of the traditional partitions are transformed by the two-storey outdoor curtain. When the curtain is closed the two storeys form one room. From inside, the space over the terrace seems to be one high closed space, so it intensifies the spatial connection between the bedrooms on the third floor and the living area on the second. At night the closed curtain has a gloomy effect. The inhabitants become fuzzy silhouettes. In this house the private dissolves, changes, transfers and is entered by the public. The frontiers between the private and the public are shifted, are replaced, overlay one another and allow ambivalence. By these qualities the building is transformed into a reflexive house even though it was realized by traditional building methods.

This was a little excursion into the developed practice of living. The three examples show how former certainties have been replaced by what Ulrich Beck and Anthony Giddens have described as self-produced uncertainty, and how doubts—or shall we say freedoms—may become form.

## References

Beck, U. (1993) *Die Erfindung des Politischen* (Frankfurt: Suhrkamp).
Beck, U. (1996) *Edition Zweite Moderne* (Frankfurt: Suhrkamp).
Beck, U. (1997) *Kinder der Freiheit* (Frankfurt: Suhrkamp).
Gorz, A. (1998) *Arbeit Zwischen Elend und Utopie* (Frankfurt: Suhrkamp).
MVRDV (1997) "Berlin Voids." *El Croquis 86*, MVRDV 1991–1997.
MVRDV (1998) "7 Notities, Bart Lootsma, Ausblick auf eine reflexive Architektur," *Arch+*, 143.

# 6 Design language

## Confluence of behavioral, social, and cultural factors

*Dietmar R. Winkler*

The Constructivist imaging vocabulary, the still accepted mainstream language for image and object-making, is a truly elitist and aristocratic visual language, with a clear hierarchical structure, made opaque by removing it from the common human experience and environment. Like all artificial languages, it has to be learned. It is jargon or expert lingo. It is not expansive. One can confine its vocabulary to a single page. It is not easily accessible to and apprehensible by the outsider. This language situates itself around a power history and the politics of the imaging worlds which exist from the times of the ancients, to the Vatican, through the epochs of nobles and the aristocracy, to today's trendsetting educated rich, stimulated by ambitious academicians, without giving up any hint of its cultural coercion.

This vocabulary includes concepts like "harmony," "balance," "quality," "truth," and "aesthetics," loaded with value ramifications. In real life there are many legitimate aesthetics, and harmonies as well as disharmonies that sometimes resolve themselves into new harmonies and aesthetics. What is disharmonious for one cluster of politics is constructive energy for the next. The Constructivist movement imbued form, physical form, with personality and character, thereby letting it emulate the human. The language is nevertheless abstract and removed from the direct human experience. In reality, the only reason we can sympathize with aggressive or passive forms is because we have experienced the pictorial abstraction through direct human physical experience in everyday life, on the bus, the subway, in a child's shoving match, or in its abstraction in the bullying of a political leader or union or corporate boss. The rewriting of the form language and its definitions through human behavioral filters would make the critique of design and art totally understandable to all participants, experts, and novices alike.

In contrast to the opaqueness of the Constructivist lingo, the Japanese language of forms is constructing a taxonomy that is direct, easily understood and easily interpreted. For example, the objects or images that respond to the concepts of "forms of stacking" or "forms of separation" do exactly that. Neither novice nor expert are confused.

Most texts written and prepared for the design profession are oblivious to the impact on the culture and sociology when a mark is made or an image or object is produced and released by the maker into a social system. That each random introduction of an object or image has the ability to enhance, but also to impede or even destroy the culture, has not yet become the central discourse of design. When it comes to communication, usually simplistic diagrams, adapted from signal and information sciences, stand in for the complexity of human communication. There is no content or context provided in describing the process of encoding, selecting the signal, transmitting, receiving, decoding the message and avoiding noise. For example, even though thermal dynamics instructs on concepts and logic of entropy and provides an analogy which makes it translatable to advertising in terms of message success through continuous hypnotic repetition, paralleling thermal dynamics' dynamism, the instructions are about the process only, not about either contents or contexts. The closest discussion dealing with social and cultural communication that one can find is in the marketing literature. This is where a concern for gender, age, ethnicity, lifestyle, even religion is entering the discussion. But marketing has its own isolated demons, namely its predatory agenda. It addresses itself to a very small portion of the rich spectrum of communication. It is usually not interested in supporting or building culture, but much more in culture manipulation.

In the overabundance of "how to do it" design books, there has been a dearth of texts that deal with the sociology of communication and language. That is why for nearly thirty-five years Edward T. Hall's writings have provided the sociological platform upon which my courses in typography and design were grafted. Hall's books are simple; actually, they are deceptively simple. They are easy reads, almost throw-aways, until one unfolds the concepts hidden in them and sees the bullion cube turn into an amazingly rich and rewarding feast. There has been nothing more exciting to students than to realize their own response to the culture that surrounds them.

The other multidisciplinary resource has been the papers presented through the *Visible Language* journal, first under the editorial leadership of Merald Wrolstad and later by Sharon Poggenpohl. From the moment of its conception in the 1960s, this journal brought together the sociological, psychological and philosophical filters through which visual language can be explored and communication design must be addressed. *Visible Language* provides one aspect: it is concerned with research and ideas that help define the unique role and properties of written language. The basic premise of the journal is that writing and reading form an autonomous system of language expression which must be defined and developed on its own terms. *Visible Language* also takes up the challenge to understand and explore the conventions for the new media, beyond the book and into more fluid and seamless systems of presentation.

Now, when global commerce and global communication challenge all cultures, the Eurocentric and design-egocentric legacies need to be regarded with great caution, particularly in view of major demographic cultural shifts, where, for example in the US, the Spanish-speaking population is increasing in an intellectual environment dominated by an Anglo-Franco-German not to forget Italian history or, on a global level, where Asian and African populations are finding their response to the history of European cultural invasions and intrusions.

Even if I stay just in the confines of my own heritage, I see questions arising that are of amazing complexity and ramification. For me the most perplexing questions are: Does the world of a German look the same as the world of a person born and raised in the US? Does living in Germany, where a citizen is guilty until proven innocent, produce different sets of behavior than in the American culture where a citizen's innocence is assured until found guilty? Does not even a quick unscientific look bear out the German behavior of the overly strong concern for the surrounding gestalt ("There they are! They observe me. They judge me.") in contrast to the American's identity ("Here I am! I do not care what you think."), using Bauhaus terms, lying in the figure and not the ground?

Another mystery is establishing the true capabilities of the visual. What is the edge between what visual language, tactile, verbal, or adumbrative language-use can achieve? Nelson Goodman, philosopher, makes clear how the visual language differs from the written or oral form. He queries: Can a picture quote a picture (Magritte, Duchamp), a symphony quote another (Stravinski, Ives) or can gestures quote other gestures (Chaplin, Graham)? Can a seascape quote the picture of a ship? To some degree some of the world's artists have answered his questions but the preponderance is still that imaging cannot achieve what visualizers promise.

We know that the advertising language is militaristic. It is about campaigns, squeezing out, wedging in, forcing out. It is about starving out the competition. It has little to do with the art director's award at the end. The art director just forms an allegiance with the copywriter and the marketer to evolve a cultural weapon to annihilate the competition.

There are many examples where visual and verbal messages are built on cultural clues and social biases but because of the tradition of separation of tasks, of writers writing and designers making images or objects, the fact of everything relating to "language," that is neither just verbal nor just visual but all inclusive, has fallen between the cracks. In agreement and grateful to Franz Boas and his contributions to determining communication the core of culture and essence of life, nevertheless I am proposing to substitute "language" for "communication" and thereby change the discussion from process to content.

The need to do so is becoming more acute, because in addition to all ancient modes of communication and traditional media, machine simulation of human behavior has been added. This requires understanding of enormous proportions. Robotics form the success of understanding the muscular and skeletal human function. Speech simulation is about the recreation of physical aspects of speech. However, the use of combinations of text, image, icon, objects, in either representation, simulation, and abstraction, to symbolize or technically explain, becomes exceedingly compounded by the fact that they are transmitted across cultures on a global scale and in rapid fashion. If one adds to this fact the sense-deprived electronic environment which takes up an enormous amount of recreational time for many, the removal of tactile and olfactory experiences, at least for this moment, and the substitution of abstract and synthetic sensory experiences create an enormous intrusion into the human culture. After centuries of slow and organic evolution we must now cope with these interruptions, even though emotionally underprepared, but commercially stimulated and stressed to the extreme. This statement should not suggest fear of emerging technologies but fear of a profession which has not prepared itself properly, and through understanding of culture, to manage and guide technology, technology which is not just the physical machinery but process, contents, and context. Designers need to be educated beyond the technical to foreshadow the impact of their work on cultural communication.

It should be obvious that the following statement is superfluous. But is it? *Homo sapiens* does not experience the world through just seeing, hearing or speaking. The body with its total sensing apparatus assesses the environmental conditions on a cathectic basis (that is, intuitive, emotional, uninformed judgement), feel good or feel bad, adumbration, making benefit or loss assessments, adjusting and readjusting. It is in the unique cyclical process in which an irritation in the gestalt of the environment stimulates the flight or fight response that is sometimes overridden by knowledge that is learned but not necessarily experienced. What is clear is that for the sake of survival what one sense cannot apprehend another must, even if the brain has to abstract the experience and translate it through another sense's memories.

*Homo sapiens* muses over what is experienced, sorting things into order so that the experience about the observed phenomenon can be legitimized. Entering first into an internal, personal discourse with the new phenomenon, then letting it grow in concentric circles, until the dialogue is externalized to include others. The observer of the phenomenon must dress up the perception of the emerging gestalt in a metaphor that advances its idea or concept. This metaphor must bridge what can be clearly understood and has been tested and found to be true in the past, across the void, to a recognition or revelation in which the new concept finds its footing. The metaphor cannot be complete as it would shut down the interpretation too early. It must allow for metaphoric expansion in the discourse to advance the emerging concept into something comprehensible. Through naming of the phenomenon and adding metaphoric descriptions the new concept enters the socialization process. When the responses from experts or public reach a certain level of critical mass, having survived challenge and scrutiny, the concept, discovery or invention is integrated into the language reservoir. In return, language is forced to adjust its taxonomies, reorganize its power structures of church and law, and evolve appropriate modes of expression. The new, emerged world-view supercedes the old. The cycle is completed just to be immediately challenged by the next dawning and interloping observation and discovery.

Discoveries in science and in the arts travel the same paths. In the same ways in which slang and underground modes of expression enter Webster's dictionary, sub-cultural icons and languages like Cubism or Constructivism become part of the total language reservoir. When totally socialized there is no longer any major resistance to the idea. The idea has joined tradition. That is why the science of relativity is built on both the evolution of human rights and the ability to think freely, supported by the philosophies of Martin Luther (1483–1546), Bacon (1561–1626), Descartes (1596–1650), Locke (1632–1704), Spinoza (1632–77), Hume (1711–76), Kant (1724–1804), Hegel (1770–1831), and Kierkegaard (1813–55) on one side, and on the other by the legacy of scientists, mathematicians, and strategists like Newton (1642–1727) and Leibniz (1646–1716). Even though we recognize in Galileo's legacy the need of fighting ignorant power with scientific reason.

*Homo sapiens* enters each situation to find it not organized until a context for the observation is superimposed. Even if observations were made under the same environmental conditions it would become very clear that organizational structures, their hierarchical arrangements and emphases, as well as interpretation and meaning, shift in relationship to the filter that is chosen. Since *Homo sapiens* has the capability to shift in rapid mind-speed from one framework of observation to the next, from the pragmatic to the symbolic, from the technical to the emotional, each context provides only one layer of the total experience.

Linguists have taught that the linear grammatical structure helps clarify the intentions of the sentence. But that clear order does not help to describe and internalize the clues of the observations of the world. If one were to traverse a forest in a new encounter, with the sense apparatus and all its receptors at the highest level of alert, one would not understand the ramifications just through the visual information streaming seamlessly into one's memory but by each sensory signal that can be taken in. Sound, olfactory signals, thermal clues are processed and filed. The adumbrative ability shields us against any possible hostility and danger. That means that former experiences become the metaphoric filter for the assessment of conditions. In terms of traversing the forest, first comes what first is encountered, the order of importance or value is found later in relationship to the tensions felt at the moment of recall. Quickly discerned gestalt, determined not aggressive, is processed quickly and pushed into the unconscious to make room for those things that need greater and more immediate consideration. Anything ambiguous, when the gestalt is not clear, and where flight and aggressive responses need to be weighed with much more and exhausting care get greater attention. Therefore, distinctions between the clustering and densities of tree species or the recognition that certain species are missing come later at the point of synthesis and retrospection. It is in retrospect that the experience is made sensible. The experience is a reservoir. It holds all ingredients for subsequent narratives. If the task of observation is about expediency, to get through the groves fast, then the narrative isolates all the issues around that filter. If the subject is about meditation, the narrative carefully selects those aspects. And if the narrative is about demons, then they will be easily conjured from the absence of certain things and conditions. Language stores each of the experience's semantic, syntactic, and pragmatic dimensions. In the design for visual communication in the various contemporary sense-limited media, the understanding of the importance of the total sensory human being, including the senses that are difficult to represent visually, is paramount.

In the repetition of life, the encounters with the same object assume various new roles in each differing context. We retain the memory of each, especially the unique moments of discovery. This moment's world, of course, is different from the next, because of the independence and accumulative effect of all experience.

The majority of traditional linguists refer to language as only that which is spoken (oral), written (verbal), or heard (aural). They also include the process of forming thought, processing, and sense-making. But most of them deal with the process. It is in the content and in the context in which a subject is communicated where the rubber meets the road. The success of the language transmission lies in the compelling cultural metaphors.

The design complexity is increasing further. Even aurality within the new and emerging technologies has a new role to play. Voice, sound, and music over stationary or dynamic image and text, is asking designers to understand language that until now has been treated through typography as silent mouthing of verbal contents. From my vantage, this is why the contemporary discussion of language should begin to include everything, from mental imaging to social and cultural ceremony and custom, symbolic use of objects and images, territoriality of concepts of rhetoric, visual aesthetics, and expressions in form of law, religion, and commerce. For example, Roland Barthes' exploration of the fashion system makes a strong argument for the language of objects. He sees all ingredients united in a sub-system of language, engaged in a visual, verbal, tactile dialogue that moves into aesthetics, value discrimination, development of social status and location in a social hierarchy, and the cultural aspects of ritual, performance, and ceremony.

Languages cannot be differentiated by establishing the incongruities between speech mannerisms or the disparate rules for each speech community. The true difference is built into the worldview of each speech community, which is content in context. It is about cultural and social perception, the measurement of values, traditions, customs and rites, the quality citizens expect and the social contract through which they anticipate rewards. Even though all subsystems support the major communicative role of the overall language umbrella, there are efficiencies in certain modes which cannot be achieved through others.

For instance, Nelson Goodman points to the difficulties in the language of art in which uniform communication and fidelity cannot be achieved, because of the extreme ambiguities and inconsistencies in usage and translation of symbols. Each object is known under a great multiplicity of labels and there are great variations in the application of these labels. Things become compounded when the same symbol is assigned different meanings in different circumstances and contexts or is complicated by the instability and ambiguity in the translation of metaphors.

Even if one looks at the language of aesthetics, one becomes quickly aware that the subject matter is a holistic study that includes the filters of the physical, philosophical, psychological, and sociological languages of art, honed by the value structure and power-history of a certain society and culture.

Umberto Eco examines the language of sports which is represented through the labels and metaphors for man, society, heroic ideals, animalism, sexuality, but also through ideas of pleasure, utility, and waste. The latter he presents as a contradiction in which caloric and intellectual waste is profoundly healthy and beneficial. In sports the physical and psychological sides of *Homo sapiens* meet up with each other. This is where recreational waste frees the individual from indispensable work or slavery. He reveals the contradictions of healthiness that can be interpreted as sick or relationships between player and voyeur, in which the observer competes seriously and intellectually in the contest through sense memory, while the players may be only marginally involved, physically or emotionally, to the consternation of the audience. He points to paralanguage and its auxiliary communication devices as far more interesting than the pejorative professional jargon in which they are framed by linguists. Paralanguage transmits simultaneously a same message by different means, allowing a certain level of clarity to emerge in the overlap and intersection of variations of tone, character of voice, and body language, each providing clues and meaning to the contextual quality of the communication such as, for instance, the speaker's age, sex, social and cultural sophistication, and the speaker's ability to add meaning through pitch, tone, accent, inflection, and tempo.

In the body language of sports, *Homo sapiens* recognizes the evolutionary journey from primate to modern man. This component of language is internalized and immediate. The clues of flight or fight alertness or being in a state of inner equilibrium are read and processed quickly. This is where the non-verbal communication mode, made up of gestures, expressions, and postures, is effective as a bridge between the organically perceived and the abstracted graphic and photographic images, that make the stone "stonier," that heighten the memory experience, that become the catalysts for the resubmission of memory from the unconscious to the forefront of consciousness as a new experience of something old, forgotten, and ignored.

Vygotsky speaks about the unity between perception, speech, and action, and the internalization of the visual field which constitutes the central subject matter for any analysis of behavior. Since perception includes all human sensory gathering-tools of stimuli, it becomes an interesting debate on where the physical brain begins and where it ends. In my book the nerves are the brain. In the same way the question of language becomes complicated. Does language only exist in the verbal and visual, or does it include the total system of sense-making and expression, message reception and processing within specific ecological, environmental, political, and social contexts?

What about the visual dialects designers own, equal to the regional varieties of a language distinguished by pronunciation, grammar, or vocabulary? There is the language of cars, the landscape, the factory. Just like Plattdeutsch—a speech related language dialect of German, or Cockney, a dialect of English, the wealth of visual expression in photographic, illustrative, and graphic varieties of language that with other varieties like cartooning, caricature, and concrete imaging, typography, cartography, topology, constitute the visual language realm of language—no single variety is standard or has the same purpose. It is the rich occupational language of art and design, and all other manners of expression which finally form culture.

There are many examples that support the notion of visual and verbal languages paralleling each other, supporting, underscoring each other. A visual language equivalent can be found for nearly each rhetorical rule. An example of interaction can be found in the design pidgin language developed for computer and human interfaces. The designer invents a simplified trade language for the sake of user-friendliness, using part of the language system of computing, data generation and data manipulation, and the language system of visual literacy and human conduct, and for expediency's sake adopts the shortest and most efficient way of each discipline to communicate. It is not difficult to find other visual equivalents for each of the language terms, glossary or axioms. Therefore, it is possible to say that architects are working with the visual interface of verbal language, in which for example the concept of "diglossia," a language phenomenon in which different dialects are spoken in different social situations, is applied. Architects understand and therefore counteract the phenomenon, otherwise banks and churches would look like Jack-in-the-Box or Kentucky Fried Chicken fast food take-outs.

This holistic view no longer allows naive discussions of dominance or inferiority of one part of the language system over another, even though one may be used more frequently than another or for different purposes or under different circumstances. The contemporary view requires that we understand the rich complexity of language as a reflection of the human spirit, existence, and evolution, not separable from culture but as culture itself. This requires all searchers to respond with humility, in awe, and with pride, because *Homo sapiens* has evolved from physical territoriality and its conflicts to social and cultural understanding. So has the design field. The next stage is tolerance, because all human endeavors are humanistic, all cultures healthy and valuable, and all subcultures ready to move up the ladder to legitimacy.

## Acknowledgements

My own contribution to today's discussion is to honor those colleagues of mine that during my teaching career have helped build the social and cultural scaffolding for the infrastructure of my course agenda.

There are a number of distinct academic colleagues at the University of Massachusetts, like Jim Hijia, Professor of History, who uses cartography to define the negligence of cultural historians in looking at the technical expression of history of Native Americans; Donna Huse, Professor of Sociology, who incorporates cultural reporting into her contribution to design classes, pointing to the conflict and clash of oral and written cultures in this century, or Judith Sims-Knight, Professor of Psychology, one of the hardiest design critics, who queries deeply the slogans of the design mega-institution, including traditional quotations like "good design communicates" or "an image is worth a thousand words." She is much more careful about these design truths. First she points to circumstances in which the axioms are definitely true but to others where visualization is just not adequate, neither dependable nor efficient. She expects designers to test their assumptions and to amass knowledge before they call themselves professionals. Then there is Richard Upchurch, Professor of Computer Engineering, who although an expert in software development, interested in human factors, questioned the cultural barriers to cyber-collaboration. He is also very concerned with what happens to the message after it has been released. What is its achievement? What is the cultural result?

The persons who really fortified my belief in social and cultural understanding are, of course, Len Singer, a very sensitive, perceptive, and extremely generous Professor of Design, committed to the cultural and social issues drowned out by present-day jargon of "human factors," and Sharon Poggenpohl, editor of *Visible Language*, who in her early academic career presented a very detailed thesis on the process of value transaction between object-maker, object, value determination, and user. She spawned my interest in the dynamics of the gift-giving culture from the Potlatch ceremony and the Hallmark greeting card to colonialism, Mother Theresa and the welfare state, the system of social and cultural relationships and obligations.

Many of these colleagues came out of their isolation and the success and safety within the walls of their own disciplines to compare notes, possibly only in the spirit of synergy, in which each discipline contributes to a shared resource of knowledge because not any single group or individual can address the rich and all encompassing field of language alone. Without these silent contributors to my philosophical platform the contents and context for my teaching would have been less strong and definitely vocational or barely professional.

# 7 Design moves

## Approximating a desired future with users

*Sharon Helmer Poggenpohl*

### First move

Nearly six years ago, I met and team-taught a communication design workshop with Chan Screven, a behavioral psychologist who is involved with visitor studies in the context of museum exhibition design.[1] Chan introduced me to quick and adaptive prototyping and user observation. We used the Field Museum of Natural History in Chicago as our site for fieldwork as we critiqued existing exhibits through watching visitor behavior and designing interventions to improve the attraction, interest and retention of exhibit information and experience. Until my work with Chan, I, like so many designers, worked in a technical and aesthetic vacuum, the standards of which were dictated by client budget and the professional norms of design organizations who tended to applaud what was "new," exotic or avant-garde. In that context, I could not tell whether what I designed communicated to its audience. The performance of what I designed was conflated with marketing strategies, the cost of competitive items, advertising agendas, and other non-design initiatives. There was no way to really know whether the message I had developed got through to its audience and altered understanding or behavior.

Chan changed my life as a designer: he showed me that we could design interactively with our audience or with the end-user of our communication; he showed me that getting out of the designer's straightjacket of technique and aesthetics by working with users could generate vital, innovative work.

For the past six years, I have taught a communication design workshop for graduate students at the Institute of Design. What we design are information products—this is in contrast to the typical communication design activity which packages predetermined messages in predefined vehicles. The focus of this workshop is the "Future of Learning." Using Chicago Public Schools as our field sites, we have been designing computer-based learning materials for middle school children. We observe class activities, identify problems and develop prototypes that we bring to the classroom so that our end-users can interact with what we are designing. Our design develops in sometimes surprising ways because of the honesty of the children and teachers who interact with our ideas.

Learning is about cognitive adaptation—understanding, transference of ideas, and integration of ideas, information, and techniques to other realms of use. While understanding may be invisible or only subtly visible, using or putting learning into action can be seen through the children's behavior and response. For the past few years, I can clearly see whether ideas are interesting or effective and how well their communicative form is working, but this has required a significant change in the design process.

Designing interactively and iteratively has changed the design process and that is what I will discuss. Rather than present a specific project, I will share with you some observations and ideas developed over the past six years of designing interactively. This will be a meta-discourse about changing the design of design.

### *Hidden moves in plain sight—what design does*

Design is a synthetic enterprise, drawing information and ideas from many disciplines. But what is more important is that it envisions the future and we all have a stake in the future. It simulates the artifacts that we desire as solutions to various problems, ideas for how to use technology in more sympathetic ways, or how to provide pleasurable and stimulating information and much more. Design envisions the future by taking a felt need or problem or what is a vague and often abstract idea and making it tangible—making it exist in the world so that the various stakeholders in the idea can imagine together, socially and interactively, what "it" might be like. They can experience the physical characteristics of the prototype and enter into a conversation based on what is empirically present before them as they explore the prototype and make critical comments or suggestions regarding its attributes. Design makes a unique contribution by focusing the future for its various stakeholders through prototype development.

Another complementary way in which design engages the future is by supporting human agency or decision-making. The common good is a search for agreement among persons based on what is collectively presumed to be good for most people. The decision regarding the common good is determined through discussion and negotiation. The prototype itself puts people in the same frame of reference, gives focus to their comments and supports their ability to come to a decision regarding the goodness or appropriateness of the artifact. With their power to create these simulations of a future, designers are in a position to negotiate in subtle ways between stakeholders as they develop alternatives with different attributes. While committed to finding the best solution, they can put the ideas of others into play and synthesize and refine the solution.

Yet another way in which design serves future development is with regard to the process of project development itself. The prototype reveals what is known or postulated for an artifact, it reveals the formal considerations and lets those involved with its development extend or reinterpret the meaning or context for the object based on an understanding of its current state. There is an ethic to this process, to not be deceitful with the prototype in terms of what is resolved, and what still provokes questions or alternatives. Disclosing all that the designer is thinking, i.e. to not hold back on controversial conceptions, hoping to slip them through at an eleventh hour, is also part of this ethic. Prototypes reveal process in a tangible way that communicates to those who are not designers the movement toward solution.

Objects and communications mediate person-to-object and even person-to-person relationships. As Philippe Gauthier notes: "objects often play an important part in the persistence or breakdown of the social bond. In short, because its definition guarantees particular interests, the object can manifest itself as either the bearer of well-being or as a vector for discord" (Gauthier 1999: 42). There are many objects that have been found retrospectively to alter person-to-person relationships. Today, for example, the influence of the computer on K-12 learning needs to be questioned for the role it plays to foster independence or technical dependence, depending on one's point of view, and how it alters the role of the teacher in the classroom.

"An ethical question underlies this same requirement for all types of intervention. To work for the common good or to give oneself the means of deliberating on the opportunism of an intervention of a technological nature requires a concerted effort in order to problematize that which is in play between persons and things" (Gauthier 1999: 44).[2] Designers need to question technology and not fall under its thrall. They need to look deeper into the consequences of technical interventions rather than take them at face value. Problematizing the relationship between people and things is exactly what designers do when they bring prototypes into the classroom for use.

There is not only an ethics to simulating the future, there is also a rhetoric built into the prototype and the language and discussion that inevitably surround it. The object seldom can speak for itself. Among stakeholders, the object or prototype becomes a subject for discussion. However, when it is before a user in an interactive setting, the prototype is presented without fanfare or elaborate context.

To summarize design's hidden moves, design envisions the future by:

- making abstract ideas tangible;
- engaging the user to help create a more fitting solution;
- focusing stakeholders on the idea as it develops;
- supporting negotiation among stakeholders as it reveals values and implications of the design;
- supporting agency or decision-making.

## Second move—the social science connection

For a considerable time, designers have accessed social science information to better develop their prototypes. This is usually information developed for other purposes that the designer believes might positively influence their design activity. Sometimes the results of these studies created for other purposes do not lend themselves to any kind of operational action. They provide a tantalizing suggestion that something may be the case, but they remain highly abstract and removed from action. This is a common problem in trying to apply scientific research in the act of designing. There is also a problem in taking qualitative research and extending it into another context.

Rather than develop a design process that relies solely on secondary research or one in which primary research is handed-off from experts as programmatic interpretations for the designer to fulfill, I have been working toward a more organic and integrated design process in which designers themselves become fieldworkers. They are observer-participants in an iterative design process that depends on user interaction with prototypes.

This organic process is predicated on several presumptions:

- designers do not need to do science in order to observe patterns and relationships;[3]
- designers need to experience and understand the context in which their design is expected to work;
- designers need to see and learn from real users interacting with their prototypes;
- the nature of design inquiry is opportunistic, alive to the moment, creative, reflective—it is not formulaic or preconceived. (This is not to say that design lacks a general process or plan, or that it is devoid of goals, but instead recognizes the dynamic nature of the development of understanding of the design problem and its possible solutions.)

This changes the design process from being immersive and abstract, largely under the designer's control, done in relative isolation to being more reflexive and tangible, subject to user participation. The iterative shift, from attending solely to the object of design followed by interaction between the user and the designed object, heightens the designer's sense of context, their ability to interrogate the developing design and to question their process and knowledge more honestly. The designer's fieldwork is transformative in terms of the development of the design. It is related to action research methods (Lewin 1951) or intervention research (Argyris 1993). Like intervention research, the design research is planned around the prototype (intervention) and responds to events unfolding between the design prototype and the user. The designer engages in cycles of action and reflection, but now the user-participant is part of the process. The goal of this iterative shift is a critical distancing from the prototype and the development of knowledge gained through pragmatic observation that is actionable.

## Third move—prototypes

Prototypes are a kind of material conversation that the designer has with the user. It is based on the idea that objects mediate our experience and understanding. This is not usability testing which seeks to verify the design of a product holistically at a rather late point in the development process, but a way to learn from the user what familiarity the object has (or lacks), what patterns of behavior the object fits into, what intuitive responses the user brings to the object and which aspects of the prototype elicit satisfaction or delight, for example. It is an ongoing conversation between the designer's prototypes and the user—it is intermittent, critical, appreciative, empirical evidence of the prototype's success or need for specific further development.

At the Institute of Design, we have identified four basic kinds of prototypes that are rarely pure in character but overlap to various degrees. They are: conceptual, behavioral, procedural and appearance prototypes.

### *Conceptual prototypes*

Conceptual prototypes externalize what are often ideas in their formative stages. In most cases, they are used to communicate ideas to individuals engaged in team activity. Because of their high level of abstraction and sometimes outright vagueness, they are not good devices with which to communicate with users. An example of a conceptual prototype can be drawn from an ongoing literacy project. It reveals search strategies in the selection of a book. This later led to the organization of search methods for downloading stories for children from the web. In this case the conceptual prototype is a diagram.

### *Behavioral prototypes*

Behavioral prototypes may differ from the actual object being designed in many ways in order to elicit from the user the behavior or understanding in question. They are designed to address particular questions and to reveal behavior through their active use. An example, also drawn from the ongoing literacy project, is a behavioral prototype created to get an idea of preferred interface through a paper representation of what might be presented on a computer screen (Figure 7.1). Students indicate their preference by touching. One "screen" has pull-tabs like a pop-up book to invite exploration and participation (Figures 7.2 and 7.3). The interactivity of a pop-up book is a good model for a behavioral prototype that simulates computer interactions. The student can write a response on acetate which can later be easily removed in preparation for another student-user (Figures 7.4 and 7.5). The paper simulation is not fancy and is cheap and easy to create. This makes it easier for the designer to avoid over-investment in the

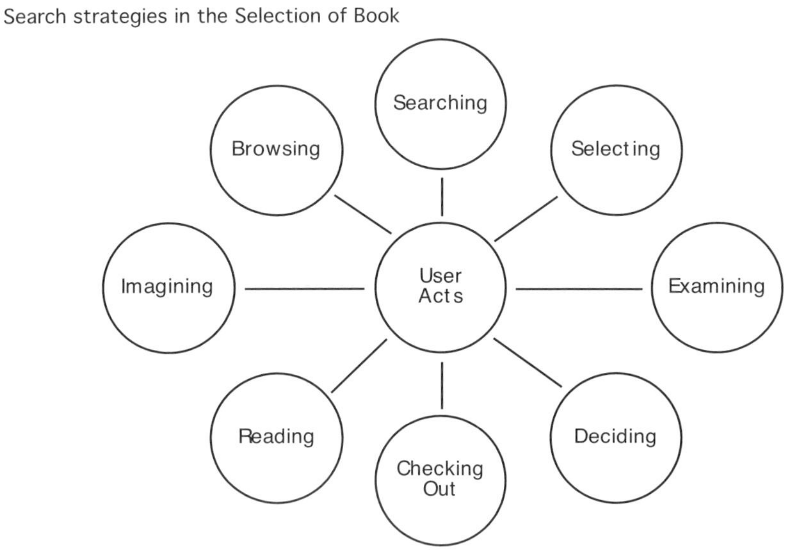


*Figure 7.1* Paper model for a computer screen.

*Figure 7.2* Inviting exploration and participation/1.

idea and allows children to respond to the idea rather than be impressed by the technology or finished appearance.

Another behavioral prototype is a computer game simulation. The object of the game is to help middle school children learn about animal adaptation. The teacher indicated that the children understood camouflage,

*Figure 7.3* Inviting exploration and participation/2.

*Figure 7.4* Use of acetate for multi-use of same model/1.

but didn't understand mimicry as a survival strategy. Children were introduced to "snakes" some of which were poisonous. Some of the non-poisonous snakes had a strong resemblance to the poisonous ones. The students were invited to test a hypothesis about how mimicry works: a predator will avoid a non-poisonous mimic that looks like a poisonous

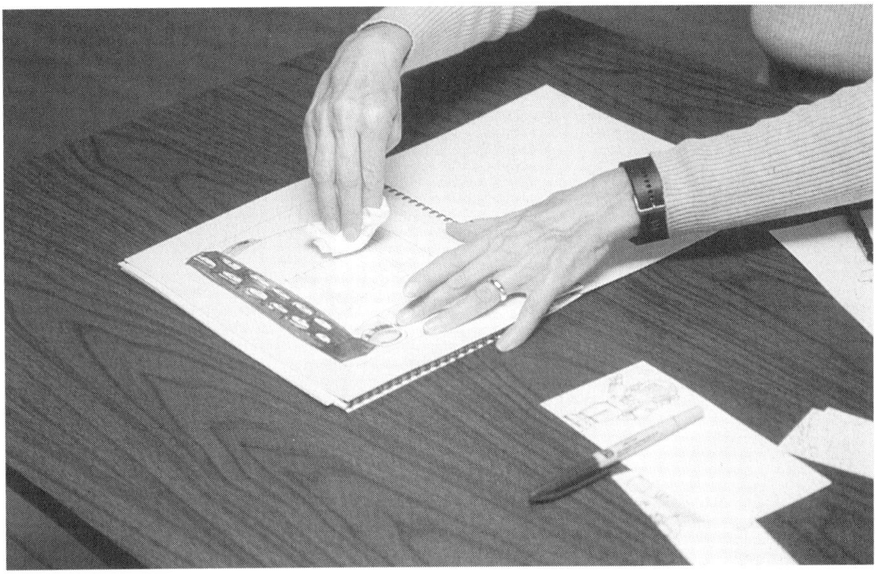

*Figure 7.5* Use of acetate for multi-use of same model/2.

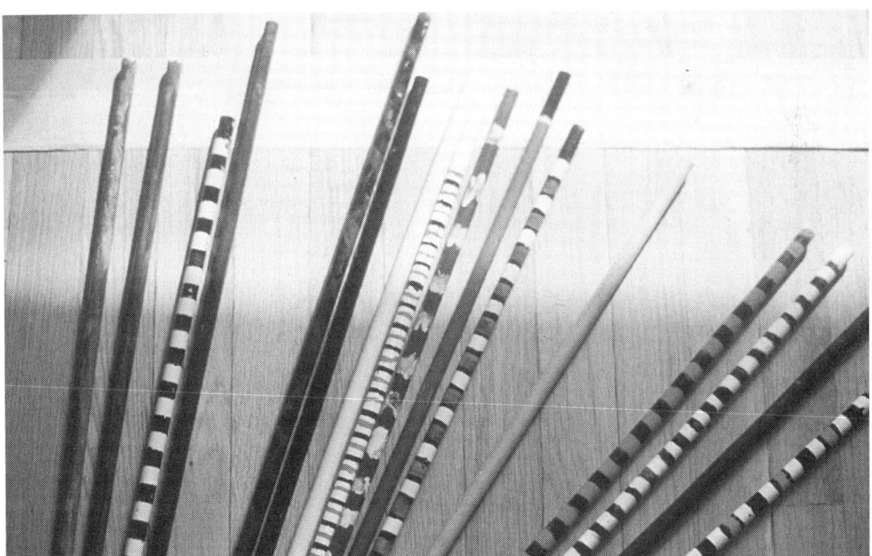

*Figure 7.6* "Snakes" for the game.

snake. The snakes were painted dowels (Figure 7.6). The children were coyotes who had to decide whether or not to eat the snake presented. They ate by stepping over a tape on the floor (Figure 7.7). Tallies were kept indicating which snakes got eaten and which children ate them (Figure 7.8). Points were gained for eating non-poisonous snakes and lost for eating poisonous ones. Fortunately the statistics from the experiment bore out the usefulness of being a mimic.

*Figure 7.7* Use of the "snakes" in the game.

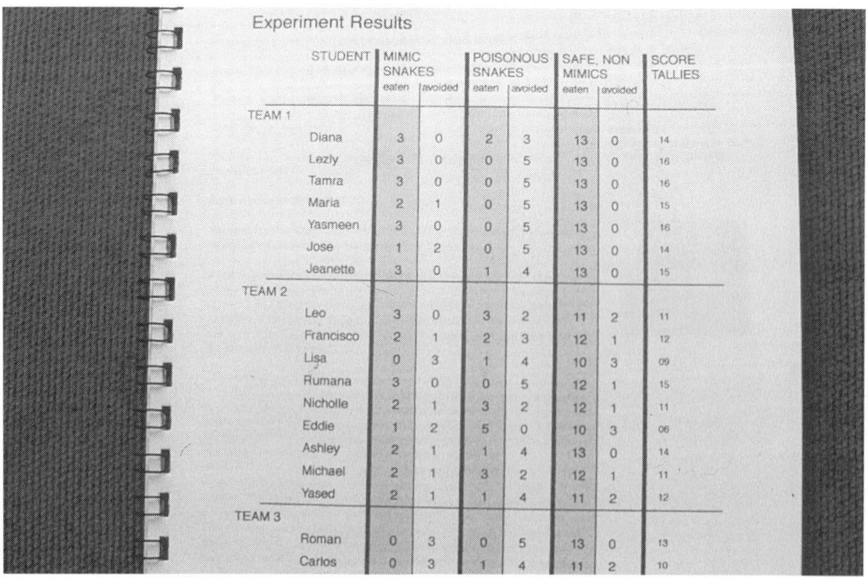

Experiment Results

| STUDENT | MIMIC SNAKES | | POISONOUS SNAKES | | SAFE, NON MIMICS | | SCORE TALLIES |
|---|---|---|---|---|---|---|---|
| | eaten | avoided | eaten | avoided | eaten | avoided | |
| TEAM 1 | | | | | | | |
| Diana | 3 | 0 | 2 | 3 | 13 | 0 | 14 |
| Lezly | 3 | 0 | 0 | 5 | 13 | 0 | 16 |
| Tamra | 3 | 0 | 0 | 5 | 13 | 0 | 16 |
| Maria | 2 | 1 | 0 | 5 | 13 | 0 | 15 |
| Yasmeen | 3 | 0 | 0 | 5 | 13 | 0 | 16 |
| Jose | 1 | 2 | 0 | 5 | 13 | 0 | 14 |
| Jeanette | 3 | 0 | 1 | 4 | 13 | 0 | 15 |
| TEAM 2 | | | | | | | |
| Leo | 3 | 0 | 3 | 2 | 11 | 2 | 11 |
| Francisco | 2 | 1 | 2 | 3 | 12 | 1 | 12 |
| Lisa | 0 | 3 | 1 | 4 | 10 | 3 | 09 |
| Rumana | 3 | 0 | 0 | 5 | 12 | 1 | 15 |
| Nicholle | 2 | 1 | 3 | 2 | 12 | 1 | 11 |
| Eddie | 1 | 2 | 5 | 0 | 10 | 3 | 06 |
| Ashley | 2 | 1 | 1 | 4 | 13 | 0 | 14 |
| Michael | 2 | 1 | 3 | 2 | 12 | 1 | 11 |
| Yased | 2 | 1 | 1 | 4 | 11 | 2 | 12 |
| TEAM 3 | | | | | | | |
| Roman | 0 | 3 | 0 | 5 | 13 | 0 | 13 |
| Carlos | 0 | 3 | 1 | 4 | 11 | 2 | 10 |

*Figure 7.8* Tallies.

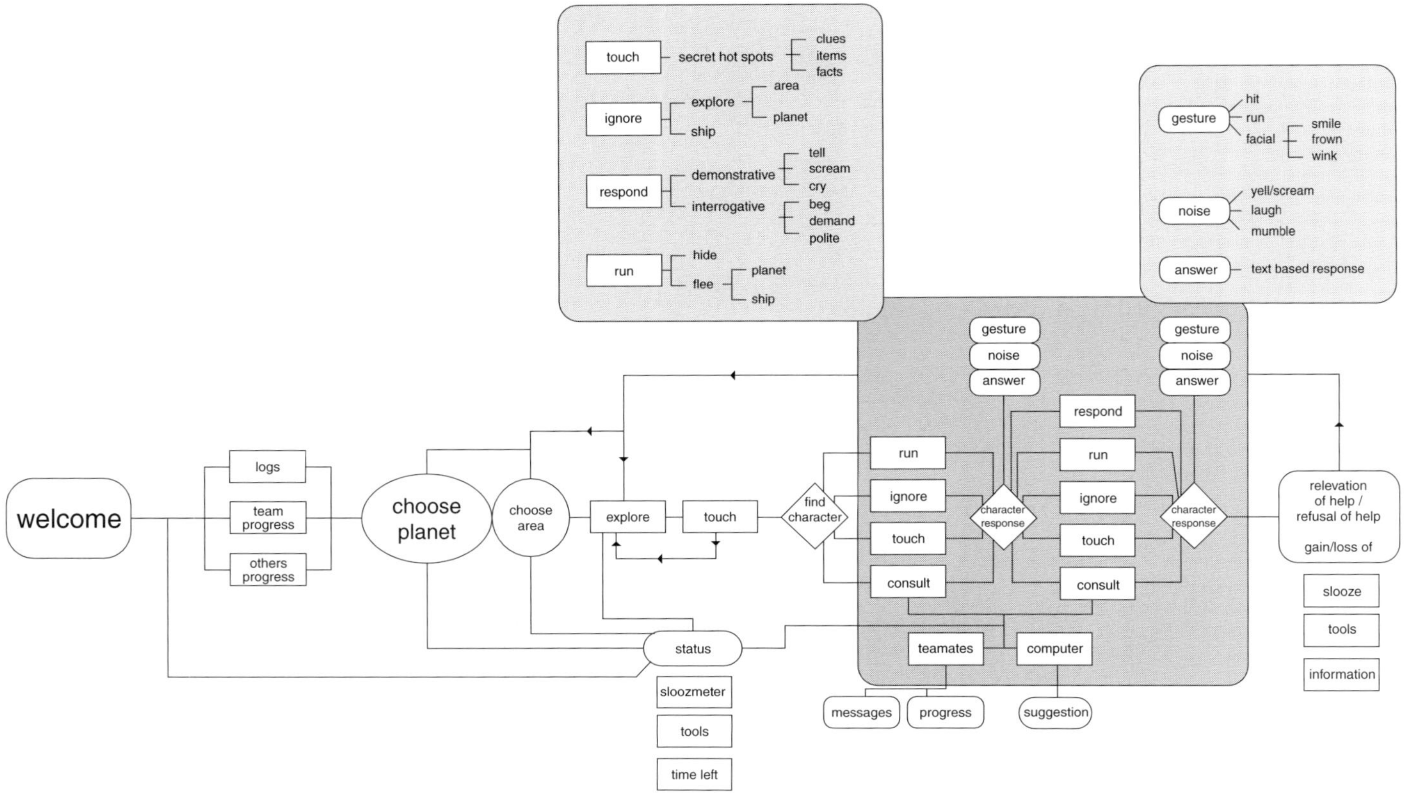


*Figure 7.9* Technical diagram of a game structure.

### *Procedural prototypes*

Procedural prototypes examine the logic of a system, to check on what is missing, confused or needlessly redundant. These prototypes can be quite abstract, are often diagrammatic in nature, are useful for communicating to project team members, or can be developed more simply and in greater detail for user investigation. An example is a technical diagram of a game structure used to communicate with the design team (Figure 7.9). In contrast, for the child-users, the game's hyperlinked story was developed as a simple paper prototype (Figures 7.10–7.12). This paper version allowed the designer to see typical paths and patterns that the children selected through the story. Both the more abstract process diagram and the more concrete hyperlinked story were useful to check logic and consistency in game development.

### *Appearance prototypes*

Appearance prototypes usually occur late in design development and are used for feedback from project team members as well as users. The qualitative character of the design is the focus: the physical characteristics (color, size, shape, weight, etc.), and the associative nature of the design (analogies, metaphors, references to other objects and experiences). The example is a computer-rendering of an electronic book for the ongoing middle school literacy project mentioned earlier (Figure 7.13). Three-dimensional prototypes are also often developed as appearance prototypes.

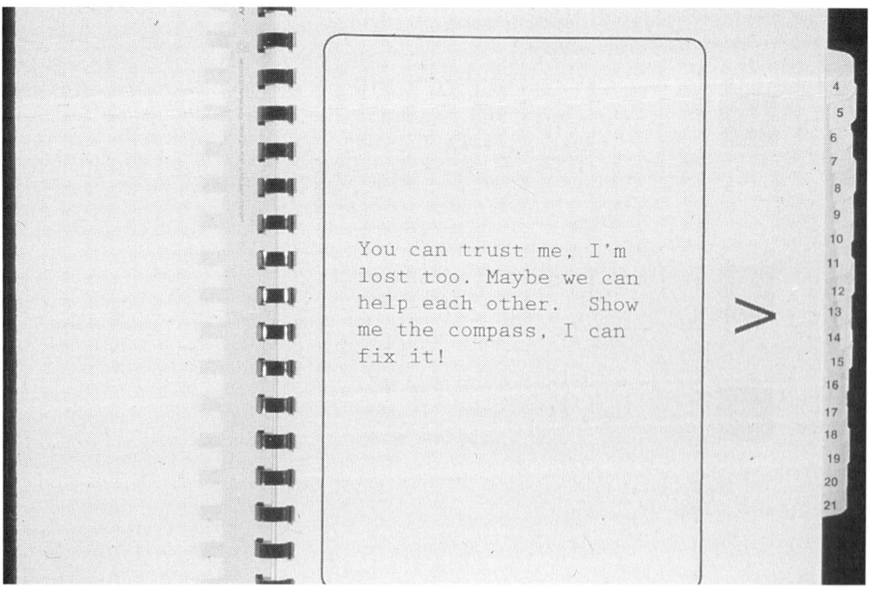


*Figure 7.10* Paper prototype for child-users/1.

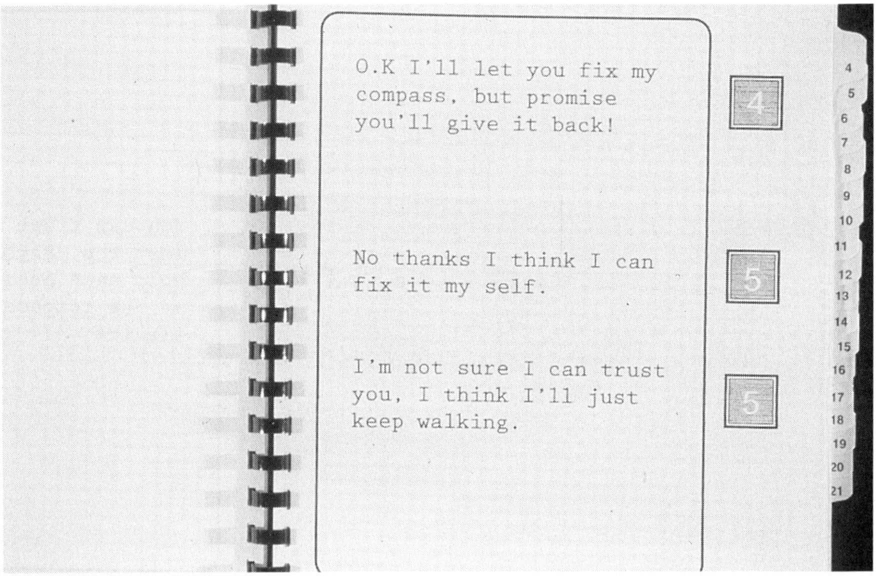


*Figure 7.11* Paper prototype for child-users/2.

*Figure 7.12* Paper prototype for child-users/3.

This is the kind of late in project development design that designers are usually associated with.

As mentioned earlier, the four kinds of prototype we are working with are rarely pure, but tend instead to overlap in various ways. Here are some common overlaps and the issues they help the designer address (Figure 7.14).

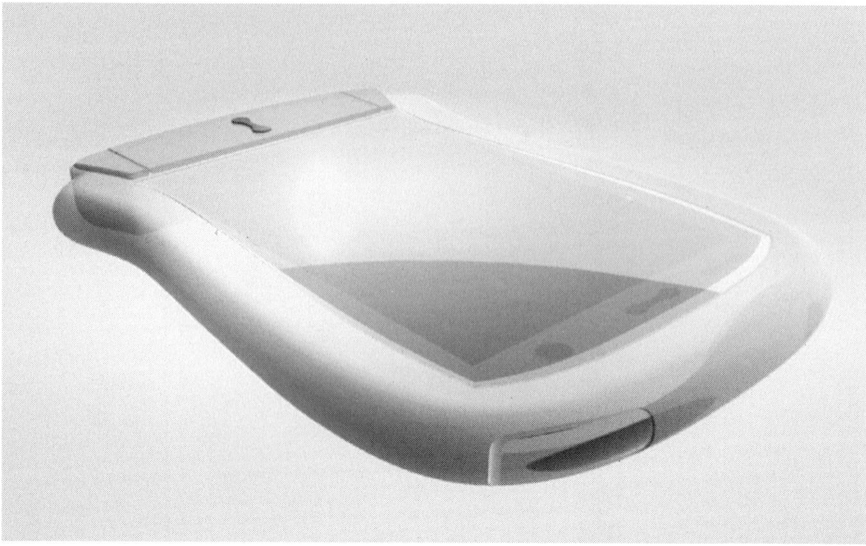

*Figure 7.13* Computer rendering of an electronic book.

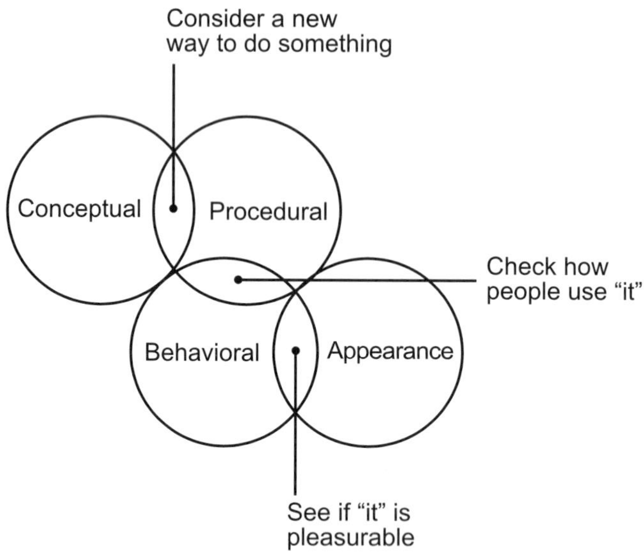


*Figure 7.14* Overlaps between four kinds of prototypes.

An overview of prototypes in terms of their typical stage in design development shows changing goals related to process and the general character of the prototype representation (Table 7.1). The observational needs also change depending on where the designer is in the design process (Table 7.2). Development questions, prototypes, and observation strategies need to fit together to sustain an iterative and interactive design process

| *Prototypes* | *Design stage* | *Goal* | *Representation* |
|---|---|---|---|
| Conceptual | very early | externalize idea | diagram, sketch |
| Behavioral | early – depends on what designer needs to know about user | elicits behavioral interaction | paper model, computer simulation |
| Procedural | middle | organize logic, sequences | space/time sequences or holistic model |
| Appearance | late | establish aesthetic, quality characteristics | refined model or interface components |

*Table 7.1* Prototypes.

| *Observation Strategies* | *Early* | *Middle* | *Late* |
|---|---|---|---|
| | watch for patterns of use or behavior | develop questions that need answers | present alternative solutions |
| | don't prejudge what is seen | develop hypotheses about user-prototype interaction | |
| | look for connections between things | | |
| *Design Stage* | open-ended problem identification | investigation – solution attempts | narrow to a solution |

*Table 7.2* Observation strategies.

that includes users, provides operational knowledge for design development, and furthers the design movement to solution.

At the Institute of Design we begin the design process with problem identification which often if not always requires observation (Figure 7.15). Discovering and developing the real problem is not trivial. The development and use of various kinds of prototypes help develop useful informa-

Process Diagram Part 1

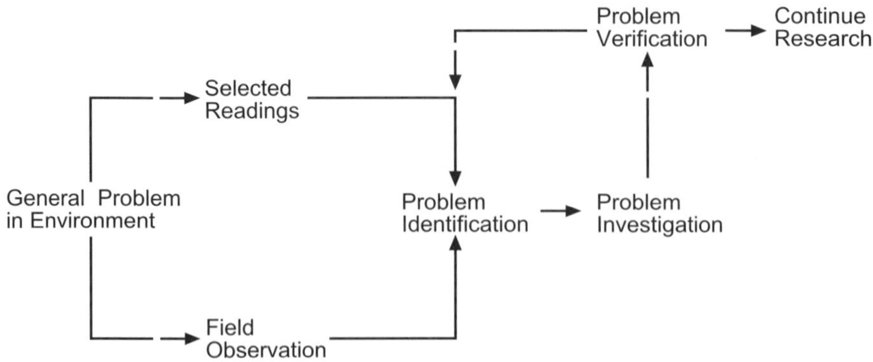


*Figure 7.15* Problem identification.

Process Diagram Part 2

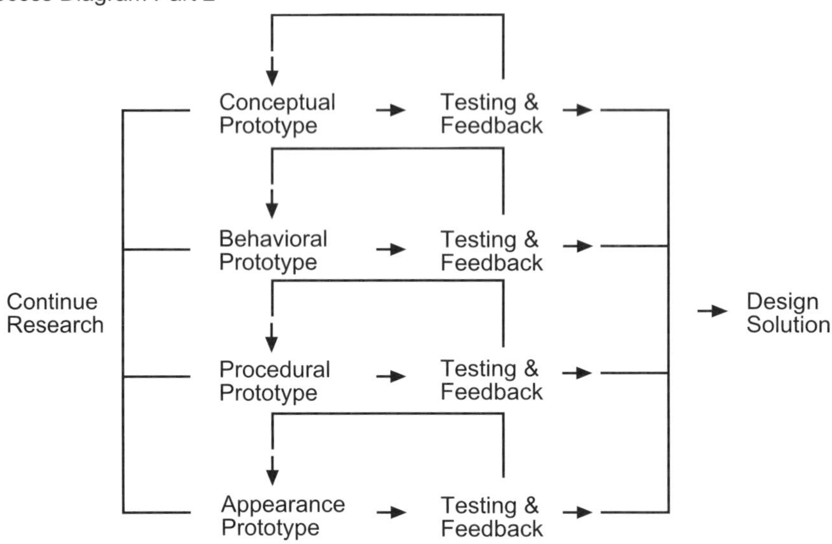


*Figure 7.16* Representation of the design solution.

tion and check on design ideas before the design team is too committed. These prototype ideas and the information they generate feed into a final prototype or representation of the design solution (Figure 7.16). This changed design process which includes the use of social science techniques and the iterative development of prototypes subject to practical use is called human-centered design at the Institute of Design, because its focus is not solely technology or economic gain, but real, practical human satisfaction with the objects and experiences designers create.

## Acknowledgements

While this article is written from my own experience, I must recognize my colleagues at the Institute of Design who contribute to the development of these ideas, the graduate students who gamely put them into play through their projects, and colleagues beyond design in the social sciences, such as Christena Nippert-Eng and Rick Robinson, who provide an interdisciplinary perspective.

## Notes

1 Chan Screven edits *ILVS Review, A Journal of Visitor Behavior*, published by Exhibit Communications Research (ECR), Inc.

2 This is not to say that design lacks a general process or plan, or that it is devoid of goals, but instead recognizes the dynamic nature of the development of understanding of the design problem and its possible solutions.

3 When science is called for, designers should work with scientists on interdisciplinary teams. For a discussion of the relationships between professions and whether design qualifies as a science, see Charles L. Owen "Design research, building the knowledge base" (*Design Studies* 19, 1: 9–20, 1998).

## References

Argyris, C. (1993) *Knowledge for Action. A Guide to Overcoming Barriers to Organizational Change* (San Francisco: Jossey-Bass Publishers).

Gauthier, P. (1999) "Technological intervention and the malady of happiness." *Design Issues*, XV(2), 40–54 .

Lewin, K. (1951) *Field Theory in Social Science* (New York: Harper and Row).

# 8 Emotion and urban experience

## Implications for design

*Ausra Burns*

### Introduction

We relate to our environments emotionally. Though design theory sometimes fails to appreciate the complexity and the variance of human experience, for anyone concerned with design, cultivating the ability to recognize, listen and respond to what people undergo and feel is vital. I intend to draw attention to the discursive topic of the urban dimension of emotion. While I focus on experience of the city, I believe that the implications of my arguments are relevant to design on many levels, as it relates to explorations of diversity of human experience in general.

Emotional experience in the urban context has been discussed in various schools of thought and within the disciplinary circles of sociology, psychology, anthropology and geography. Certain prominent themes and representations of emotional reactions to city life enrich the interdisciplinary dialogue and broaden our understanding of issues and modes of conceptualizing the contemporary urban condition.

My intention is to bring into design discussions centered on people's experience of, and reactions to, built environments, certain aspects of disciplinary knowledge arising from the social sciences and humanities. I will communicate some of the conceptual discourse surrounding my theme toward revealing dialogical, and more holistic, context sensitive ways of designing for the city—ways that use effectively knowledge generated in other fields of research, and accordingly create paradigms for design action.

There are many ways to approach and discuss the diversity of the city. Along with the transformations of modern history, and the growth of Western metropolitan culture, visions of the city have come to be divided into spheres of disciplinary competence. The complex phenomenon of the city is often defined by architects as a depository of building styles and influences, by economists as a site for regularization of retailing practices, and by planners as a transportation node and mosaic of municipal bylaws. On the other hand, a single hegemonic conceptual framework will not be able to account for and recognize all the diversity and conflicting notions of urban culture, its forms and social processes. Consequently, it can be argued that no one possesses all of the knowledge and wisdom required to understand and act responsibly in this world. "We need diversity and alternative perspectives to keep alive the ongoing inquiry into ordering, disordering, and reordering that is the central enterprise of human culture" (Buchanan 1998: 15).

The changes experienced in contemporary Western cities over the past several decades—gentrification, suburban sprawl, physical and social fragmentation of the city, its growing cultural and physical diversity—have been identified as dominant developmental trends. Searching for a narrative about the city and its economy "that includes rather than evicts" marginal economies and their representations in the city's physical form, political economist Saskia Sassen establishes an intellectual interdisciplinary dialogue on the subject of race, gender and representation in the city (Sassen 1996: 184). In so doing, she provides us with an example of theorizing in the critical space in between disciplines, where new knowledge and directions for constructive action can be fostered. The need to embrace new models of reflection on urban processes has in fact been voiced by many social theorists and practitioners.

"Emotions" are the complex conjunction of physiological arousal, perceptual mechanisms, and interpretive processes; they are thus situated at the threshold where the noncultural is encoded in culture, where body, cognition and culture converge and merge (Illouz 1997: 3).

Why do I want to induce discussion on such a relativistic and seemingly unscientific subject related to urban experiences and emotions? My answer is that I think the potential benefits of this strategy outweigh the risks in terms of creating a better understanding of the phenomenon in question and developing design strategies that work in specific socio-physical conditions. Through discussion of how we or others feel about city living—in various situations and differing circumstances—we can increase our ability to relate our aesthetic responses to perceiving the environment to the practical actions we take within it. In terms of study of the subject and of theorizing, such an exchange may present us with the opportunity to critically analyze the standpoint of the researcher as a detached observer. In forums of discussion on how individuals or certain urban subcultures perceive their lives in the city, the researcher may become a more involved participant whose own stake in the issues at hand is raised and uncovered. Inevitably, when designers become more involved in the issues that concern people affected by design changes, they open themselves up to professional scrutiny and the challenges of self-definition. In order to follow and refine such a direction in professional and intellectual practice, one must identify the ontological, epistemological and methodological aspects embedded in one's theoretical orientation. Questions that could be asked in this context include: Whose interests are represented in the project? How are the results of the research or the physical changes to an environment going to impact on various parties that share the common urban realm? Questions could also extend beyond the immediate concerns of the designer. For example: What will be the impact on future transformations of decisions made today?

In the context of the city, the experiential realm is largely comprised of the places and objects of everyday life. Streets and backyards, parks and monuments become situated not only in the realms of architecture or urban planning, but also in the realm of the human environment, where a distinct object, feature or image is dissolved "into a world of perceptual experience" and can no longer be regarded as an "external location but as continuous with human life" (Berleant 1991: 77). In this way, broader cultural aspects of the formation of emotional responses are grasped through research that goes beyond the study of the physical qualities of urban form: its colors, smells, forms and textures. People's emotional relationships with the environment are framed by attributes that may be evident from an external assessment of a situation—for example, skin color, class, gender and social status—and by those attributes that are not evident—those formed by individuals' personal histories and life events.

This position may represent an alternative to the traditional modes of framing experience in the designed urban environment. The tradition has been to view urban form as the static, axially oriented visual space of Western Classicism. Throughout Western social thought emotions are seen to be the very antithesis of the detached scientific mind and its quest for "objectivity." The roots of this separation and the custom of repudiating the importance and integrity of emotional experience lie deep in the Western intellectual tradition that separates body from mind, nature from culture, reason from emotion, and public from private. Moreover, these dichotomies are not value-free. The hierarchy intends to establish the supremacy of reason. Progress and precision are held above emotional, private, subjective experience. This experience is located in the realm of urban spaces and associated with femininity and irrationality.

## Urban experience, its conceptualizations and representations

More reflective, culturally rich and dynamic articulations of our emotional relationship to the urban environment have been emerging from various disciplines of the social sciences and humanities for a number of years. An interest in the complexity of the social world and the positionality of knowledge and experience has evolved through the work of contemporary philosophers, geographers, anthropologists, psychologists and sociologists as part of a direction in social theory. Its aim is to develop an alternative to the positivistic and progressive view of the world as moving forward, in a linear trajectory toward . . . well, no one knows where anymore. Complementary calls have been made for radical transformations in the way we understand design as a professional activity. These calls come from those who believe that the models and modes of action exploited earlier in this century are no longer adequate. Designers and other professionals dealing with the production and modification of urban spaces, processes and imagery are looking for new ways to approach culture as defined by conflicting values (Buchanan 1998: 19). To comprehend the "dynamics of individual and social behavior well enough to work efficiently and effectively in interdisciplinary teams" requires that design practitioners and theorists seek some common understanding of the social and cultural issues at stake and deepen their awareness of contemporary intellectual discourses and research methods that can contribute to bringing design, the humanities and social sciences together (Frascara 2000:120).

The humanist traditions in philosophy and the social sciences have introduced the notion of emotional and experiential complexity in the perception of one's surrounding world, and have challenged the more traditional model based on contemplative knowledge accumulation through passive spectatorship, objectivity and rationality. Symbolic interactionists suggest that individuals are not just aware of their place in the world, they are also involved in group interactions, each of which is located within a particular social setting worthy of careful consideration (Mead 1934). Many significant works produced within this theoretical orientation recognize the value of situated and reflective knowledge. Describing ways in which the space of inner-city Philadelphia was given meaning and made legible by street gangs through territorial boundary markers, David Ley shows how the realities of everyday life are negotiated by people in concrete contexts (Ley 1981). From this study of city gangs and their territorial behavior, Ley infers that the space, and the emotional responses to it, are socially constructed. The sociological study of the partitioning of the city into numerous territorial worlds has been a major aspect of the Chicago School tradition. In the works of members of this school, the impact of the overall diversity of the city on urban dwellers is much reduced. The main arguments comprising the standpoint of the school rest on the assumption that the city is a mosaic of different social worlds which overlap and interact. People, therefore, create their own social and territorial niches in the city, and in this way are able to develop a sense of identity and comfort in the modern metropolis (Langer 1984).

Attempts have also been made to articulate the conception of experience through close association of the space and the perceiving body. From this perspective, the environment that is perceived does not exist solely outside of the perceiver. It extends the "inner landscape of human beings into the world in ways that are comprehensible, experiential, and inhabitable" (Dewey 1958; Merleau-Ponty 1962; Bloomer and Moore 1977: 105). The "active model" oriented on action, function and response to one's surroundings has been developed by the American pragmatic tradition and in European existential–phenomenological philosophy (Berleant 1991: 87). Humanist geographers sought, through the philosophies of existentialism and phenomenology, to recover the essence of the experience of place. They propelled the shift toward recognizing the materiality of everyday life and the power relations that influence the emotional reaction to place. Nevertheless, as critics note, the humanists still have not come to terms with the depths of subjectivity and intersubjectivity (Pile 1996: 62). Another serious criticism comes from feminist geographers who remark on the lack of adequate theorizing on the broader social power relations structuring our experiences of place. One of the underlying reasons for this deficiency is that humanists assume masculinity as the implicit norm through a certain form of rationality that still considers objectivity as the touchstone of true knowledge. In a 1984 meeting of feminist geographers, it was argued that "humanists tend to show a general concern for the way in which ordinary people are subject to various forms of authority, rather than analyzing the specific forms of exploitation and oppression that occur" (Rose 1993: 44).

Turning to linguistic philosophy, we learn that the emotional realm is woven into the structure of communicative action, and therefore open to contestation and argument. Intersubjective structuring, or communication between individuals of their subjective sensations through verbal and body language, becomes possible because emotion is publicly observable. It takes the form of actions made in response to certain circumstances. In other words, emotions can be conceived as meaningful responses to life situations (Crozier 1994: 19).

Some radical philosophers such as Marxists argue that people travel through a time–space life path while internalizing and interiorizing social relations. Human agency must be framed not only within the determinations (or power relations) of social structure, but within the material properties of time–space relations, and within the processes inherent in "personality." Personality in this case signifies identity that is expressed through subjective reactions conditioned by the life history of each individual (Thrift 1996).

In the work of behavioral geographers, mental processes and cognitive representation are of central importance. A more thorough understanding of human cognition was critical to establishing links between the mind and behavior. Mere descriptions of overt patterns of behavior were replaced by a search for, and explanation of, the reasons why people behaved in certain ways. But, as recent critics note, behavioral geographers failed to "recognize the mutual interaction between mind and environment" because they still operated in the realm of dichotomies such as those between the external and internal worlds, between the public world and the private world, between the subjective (perceptual) and the objective (phenomenal) world, and between Mind and Nature (Pile 1996: 43).

Exploring different spaces of the contemporary city, feminists often reject the pursuit of generalizations and "complete" visions. Their work is more tentative. It is grounded in the details of the everyday, and enables interpretation of social life and spaces in the city as heterogeneous (Rose 1993: 133). The strategies through which feminist geographers pursue their goals include undoing, subverting and transcending the power-infused dualisms between dynamic and progressive time and static space; between the public and the private realm; between rational knowledge of, and the emotional responses to, the environment; and hierarchical dichotomies built on notions of masculinity and femininity. The "disorder" of urban life does not disturb women. The "socialization of women renders them less dependent on duality and opposition" (Wilson 1991:282). To many women, urban spaces simultaneously represent delight, a site of connection, and a place of danger and oppression—spaces that are lived, experienced and felt.

Another important contribution to more reflexive and reflective theorizing about emotional responses to the city comes from geographers using psychoanalysis to reconceptualize the dialectics between the subject, society and space. Each individual may be seen as "tied by the bonds of love and hate, in many directions, to numerous groups; each forms their sense of self in relation to different models of behavior; each has a share of many group identities" (Pile 1996: 118). These diverse relations are spatial, but originality and richness of experience, and the strength of ties between the individual and the environment, come from within, from the agent.

All of the nineteenth-century founders of sociology touched on the topic of emotion. Among them was Max Weber, who wrote about the anxious spirit of capitalism that evolves in the modern metropolis, and the role of rationality and charisma in the formation of this new way of living and being a member of a capitalist society. Karl Marx developed a view of alienation as an inevitable consequence of class conflict that brought to Western urban centers resentment and anger toward capitalist exploitation. Georg Simmel believed that the emotional state of the modern individual being was profoundly shaped by a continuous bombardment of the stimuli of urban life (Simmel 1955). Among the most evident reactions he points to is the "reserve" attitude one develops in order to survive in the saturated life of the city. This reserve attitude remains central in contemporary discussions regarding the reactions of the postmodern individual to the commodification of culture and her or his involvement in the collective consumption of fetishized commodities.

The attempt to establish scientific legitimacy in the discipline of sociology turned many researchers in urban studies toward explorations of social action, rather than of peoples' perceptions or other "soft" images of the city (Langer 1984: 198). Despite this trend, Robert Park, a key member of the Chicago School, wrote: "The city is . . . a state of mind, a body of customs and traditions, and of the organized attitudes and sentiments that inhere in these customs and are transmitted with this tradition. The city is not, in other words, merely a physical mechanism and an artificial construction. It is involved in the vital processes of the people who compose it" (Park *et al.* 1996: 1).

From the perspective of the environmental psychologists, our rapid and largely unconscious decision-making process is influenced by the potential for functioning in the locale. Such pointers for potential functioning are perceived abilities: the ability to enter the setting, to acquire the necessary information about the environment or setting, and to maintain one's orientation (Kaplan 1989: 174–5). Acknowledgement of the complexity of human emotions and their variance depending on the particularities of individual circumstances and cultural settings has brought psychologists' discussions closer to an understanding that "it is a person's experience of the world rather than the world's objective properties that counts" (Crozier 1994: 75). Neisser's seminal book *Cognition and Reality* marked a transition for psychologists (Neisser 1976). This more reflective and holistic conceptualization of psychological responses to one's environment caused psychologists to respond to, and gain interest in, research on cognition and the mental processes that underlie behavior. It has seemed to psychologists that "physiological processes, including variations in arousal levels, are not in themselves sufficient to discriminate between emotions, but that cognitions, beliefs or attributions are also necessary" (Crozier 1994: 19). Anthropology, on the other hand, embraced emotion in terms of how its conception and expression were subject to cultural production. The question facing anthropologists today is how best to integrate the subjectivity of those they observe into their analysis: in other words, how to redefine the conditions of representativeness to take into account the renewed status of the individual in our societies (Augé 1995). "The notion of material culture, developed by anthropology, initially due to the need to reconstruct social life through an analysis of extant objects, provides a conceptual frame for the understanding of how cultural models are promoted by material objects" (Frascara 2000: 124).

## Urban experience: themes and representations

As my topic revolves around people's emotional experiences of the city, I would like to present a range of emotional responses and sensations that theoreticians from diverse fields of knowledge have identified, and later discuss how a broader knowledge of these themes and ideas can help designers in their professional work. While not representing a complete or exhaustive review of the subject, such discussion might lead designers to revisit or re-evaluate our paradigms of action and theorizing.

One of the city's strongest aesthetic appeals is to the person as pedestrian, and "this appeal rests very much on its attraction to the moving body, its ability to entice one to follow along a street in relaxed and irregular rhythms" (Berleant 1991: 101). Information derived from anthropology and psychology can support the argument that people enjoy "crooked streets" and the richness of urban experience, but they are most afraid of being lost. The intensity of this fear of being lost, disoriented or confused by the monotony of the city suggests that designers (and urban planners and architects) should strive to produce "imageable" urban space by sufficient knowledge and through conscious manipulation (Lynch 1981).

One of the most powerful and brilliant descriptions of the emotional experience of individuals roaming the nineteenth-century Western metropolis was presented by the social theorist and philosopher Walter Benjamin. Employing techniques of surrealism as well as avant-garde montage and cinema, Benjamin created the portrait of an urban drifter, the *flâneur*, whose daily experiences were embedded in the "novel kind of beauty in the streets," through mundane activities of shopping, strolling and socializing (Wilson 1991: 280). The Paris *flâneur* was lured by the magnetism of the city streets, by the sensual power of crowds, by the erotic pleasures of window-shopping and offerings of sexual pleasures outside of the family circle. One of the important aspects of urban culture that Benjamin was able to relate through his narrative, and that still remains important in the contemporary city, was that experience was atrophying—that there was a rise of spectacle and spectatorship, and that interpersonal relationships were being replaced with the packaged messages of a commodified culture of spectacle and merchandising. Simmel argues that the individual is constantly presented with myriad possibilities in the dense and varied realm of the city. The person is continually confronted with strangers, and this makes it impossible to establish any deep personal relationships. The only reasonable reaction to this situation is the adoption of a posture of "reserve" and impersonality. The Chicago School sociologists stressed another aspect of urban experience. People's ties, relationships and attachment to their particular territorial niches in the metropolis were significant because it is in these niches that they "come to have some control," where they "are able to develop the sense of identity and comfortableness that one large downtown world makes impossible" (Langer 1984: 108).

If we adopt a perspective that recognizes the city as a site of power struggles, and therefore a site of diverse and situated experiences, we can see why it may matter "who is walking the streets and who is doing the looking, and why, it also matters which streets are being walked, and how the spatial regime of the visual is constituted" (Pile 1996: 231). From the viewpoint of feminist geographers, the gaze of the urban drifter, as presented in literature concerned with issues of urban experience, is often accepted as universal, and is in fact a masculinist gaze embodying a relationship of an active onlooker and a passive object (Rose 1993: 104). From this perspective, Benjamin's *flâneur* appears to be in such a position of power. He is captivated by the movement and excitement of the urban modern, but out of fear holds to a safe proximity or distance. What stands beyond this distance is an "uncharted territory: women, masses, the city: a territory which was simultaneously psychic, bodily, spatial and social; simultaneously real, imagined and symbolic" (Pile 1996: 209).

Let us turn for a moment and look at the city through the eyes of those who are being watched. Women often feel vulnerable in public because they are seen as properly belonging to the domestic sphere (Valentine 1992). "Being in space is not easy. Indeed, at its worst this feeling results in a desire to make ourselves absent from space; it can mean that 'we acquiesce in being made invisible, in our occupying no space. We participate in our own erasure'" (Johnston quoted in Rose 1993: 143). Rose recalls her personal emotions about being in everyday spaces of the city: "I have a strong sense of space as oppressive, for example, from being scared walking at night in the city in which I live" (Rose 1993: 143). On the other hand, the city's crowds and spaces make it easier for many men and women to become anonymous, to escape to a certain degree from the control of traditional hierarchies.

Power relations and their symbols are embodied not only in the actions and relationships of city dwellers, but in the spatial forms of the city, within the anchoring points of its architecture. The power of authority is displayed in the centrally located skyscrapers of contemporary cities, often housing the dominant economic, political and state power of the city. One of the important issues here is that this spatial organization seems to give an impression of intelligibility and transparency. Modern architecture's abstract transparency alludes to the utopian vision of a "radiant, egalitarian, dynamically open society" while embodying the "reality of panoptic, hierarchical bureaucracy" (Ockman 1996: 205). What is lost or, more accurately, pushed away or erased, is the representation and acknowledgement of the subordinate, marginalized, less powerful cultures inhabiting the urban realm.

Where then, is this marginality of the city embodied? How can we "excavate" this experience of the "other"? "Otherness" is embodied in the places of everyday: the homes, parks and shantytowns of our cities. There is no homogeneous "other" behind this theoretical cliché of "otherness." There are actual, flesh-and-blood others (McLeod 1996: 21). Some of them the sick, disabled and elderly, find a certain degree of comfort, security, autonomy and even freedom in sites of everyday life such as the home, the public park, and the department store. While the home, under certain circumstances, to some people, can be a source and site of oppression and violence (women, children in abusive families); to others, and in other circumstances, it can be a place that fulfills deep yearnings for empowerment and control over one's life and place of community contact (Hermanuz 1996: 235).

Many of us have experienced the devastating feelings of estrangement from one's surroundings engendered in such places as shopping malls, large hotels and transit points—places controlled by computer networks and personal credit-card identification. These places of supermodernity erase senses of real communication and memory. The individual becomes a passenger, customer or driver who is "possessed" by the "passive joys of identity-loss, and the more active pleasure of role-playing" (Augé 1995: 103). But despite our worries about the changes that digital technology has brought to our cities, we have to face the fact that boundaries are becoming blurred between the social and the technological, between the natural and the artificial. The important question that is asked by many is: "What are the implications for human emotional experience of these new forms of technology and the various 'hyperrealities' they spawn?" (Williams 1998: 120). It can be argued that new forms of emotional intimacy, sharing and meaning are beginning to open up as a consequence of these technological developments. "The computer network provides opportunities to get together with considerable personal intimacy and proximity without the physical limitations of geography, time zones or conspicuous social status" (Williams 1998: 124). Though the "intrusion of commerce and sophisticated technology into every crevice of daily life can hardly be considered cause for comfort, it is also the case that the built representations of postmodern society are no longer charged so heavily with dichotomous gender stereotypes" (Ockman 1996: 208).

One of the themes that recurs in discussions about feelings and perceptions of the built environment is that of diversity and fluidity, of emotion's dependency on context. Physical and virtual places appear to coexist and blend in the contemporary metropolis. This demands from designers, architects and planners significant revision of design strategies and methods of work. Such changes involve recognizing the presence of marginality in social life and formal representation of the city, and working toward recovering the "informal life" in the city's dominant representations. We need a radically new approach to cities if we are to see realized the city's potential to offer freedom and autonomy to all individuals and groups (Pile 1996: 283). In many cases, the recovery of marginality offers passages into mysterious human nature, and at the same time provides confused "form givers" with some constructive understanding of what kind of city it is that people need these days, of how that city is supposed to serve their needs and reflect their emotional yearnings in a just and responsible way (Cline 1997: 14). Mere formal subversion of "otherness" in buildings as objects of art, and placing dominant value primarily on the physical features of the designs, is not sufficient or effective toward making contemporary cities better places to live. To adopt the position of an informed designer one must pursue a deeper understanding and appreciation of the cultural issues and practices that converge in the body of the contemporary city. One of our valuable resources of creativity and skill is our ability to listen and learn from those for whom we design (Frascara 2000). In the urban context, we should not forget that public space is the representation of a "public, as a living, acting, and self-determining community" (Torre 1996: 249).

As we develop a more profound sense of how different people live in the same city, we might not be surprised to find that "the public realm that can be some people's heaven can be other people's hell" (Cline 1997: 53). I would argue that professionals should become more prepared to accept and learn from the unexpected twists and turns of real life city events. Many spontaneous and informal practices are deeply significant to the engaged individuals and groups, and can be very revealing to such professionals as designers. An example of such a practice is that of Latin American dwellers of the Bronx constructing, in the "vacant" land between apartment buildings, little houses that remind them of their home country. These *casitas* are filled on summer nights with the bustle of people enjoying comradery and the night (Cline 1997: 21). This example also reminds us of the fact that a sense of security and enjoyment of public spaces such as a park or a street depends on matters that extend beyond demands for sufficient lighting or smooth paving. Many ideas that transform our lives and are of greatest significance derive from non-architectural sources. Don't we often find ourselves perceiving the city, our daily journeys, and our inner thoughts and routines as one inseparable physical–emotional–mental landscape? Aren't our impressions of city sites and experiences of sightseeing subordinated to our daily worries and thoughts as we walk or drive along familiar streets?

While this does not mean that many traditional disciplinary skills and artistic imagination are obsolete to the designer of urban spaces, images or products, it does suggest that the issues relevant to a designer's professional competence require serious scrutiny and expansion. One of the critical aspects of this revised picture would involve the transcendence of static and oppressive dichotomies between male and female, between reason and emotion, and between the rational and the subjective. It would mean that space and time should be redefined, and seen as interrelated. "We need to conceptualize space as constructed out of interrelations, as the simultaneous coexistence of social interrelation and interactions at all spatial scales, from the most local level to the most global" (Massey 1994: 264).

Along with accepting the fluidity and diversity of concepts we operate with in our daily practice, it is necessary to adopt a flexibility in our methods of work and research. As architect Denise Scott Brown remarks, it is a sense of professional responsibility that moves her to accept the diversity and temporality of social agendas and meanings attached to designed spaces and buildings in the city. Instead of following the rigid directives of dominant ideologies, Scott Brown, in her daily work, chooses the more difficult route of negotiating mutually accepted agreements between parties involved. She admits, however, that "ideologies come and go and functional needs change with time, yet our buildings may remain" (Scott Brown 1996: 215). As difficult as it is to concede, we might never be able to determine the ultimate method or the perfect methodological package that would free us from continuous self-questioning, end our creative search, or address all the transformations in our working context. Cities, places, societies, and the emotional responses of people will change.

"Each situation demands specific responses, and all that methods can do is help us approach each new situation with a more sensitive and efficient eye" (Frascara 2000: 120).

## Urban experience and the design practice

As I have been exploring here the role of emotion in urban life, I hope that the reader has been encouraged to adopt a critical position with regard to the information and the complementary or competing ideas presented. As I asserted earlier, there is no one and correct answer to all our problems of urban living, and therefore, there is not one set of criteria according to which the design process, under the current conditions, is supposed to develop.

Nevertheless, I argue that through informed selection and thoughtful consideration of some leading ideas offered by philosophers, geographers, sociologists and anthropologists, designers may be encouraged to reconsider the paradigms they traditionally identify with. Beyond adopting greater moral responsibility for their actions, a more profound awareness of contemporary social conditions, and enhancing their knowledge of methods, designers can embrace the diversity of human experience, fundamentally shifting their standpoint within the profession. The key to making this transition, I believe, is to adopt theoretical and action paradigms that enable designers to influence social change through interpretation and negotiation. This stands in marked contrast to a design practice based on authorship and the imposition of opinions and expertise.

The diversity of these concerns does not signify to me a loss of direction in design, nor a retreat to formal experimentations. As a designer working on urban design issues, the complexity I discover in conceptualizations and perceptions of the city serves to encourage me to continuously revise and adapt my working methods to the contextual criteria of each design situation, and to assess the forces shaping people's attitudes and actions. In this context then, I contend that knowledge gained through work on design projects does not accumulate in an absolute sense; rather, it transforms us and leads us to more informed insights. Since this approach is linked to postmodern paradigms, it also can be defined as inherent in the wisdom of everyday living. Isn't this the way we, as human beings, gain life experience and life skills: moving from one experience to the next, from one life lesson to another? What accumulates, of course, is not a catalogue of events defined by frequency or location of occurrence, but images, sensations and perceptions of the critical links and relationships. This accumulation becomes a wide web of practical knowledge which cannot easily be labeled as true or false. Knowledge can only be revealed through an understanding of the cultural discourses within which it is embedded. In such an approach, whether the research and design transformations are concerned with peer relationships, work strategies, perceptions of images, or patterns of use in built environments, detached reflection based on a purely theoretical way of thinking is transformed into discussions and negotiations between the researcher and the user-participant.

## Note

1 A similar version of this paper has been published as "Emotion and urban experience: implications for design," in *Design Issues*, MIT, 16(3), Autumn 2000, 67–79.

## References


Augé, M. (1995) *Non-Places: Introduction to an Anthropology of Supermodernity* (London: Verso).

Berleant, A. (1991) *Art and Engagement* (Philadelphia: Temple University Press).

Bloomer, K. C. and Moore, C. W. (1977) *Memory and Architecture* (New Haven: Yale University Press).

Buchanan, R. (1998) "Branzi's dilemma: design in contemporary culture." *Design Issues*, 14(1) (Winter), 3–18.

Cline, A. (1997) *A Hut of One's Own: Life Outside the Circle of Architecture* (Cambridge, Mass.: MIT Press).

Crozier, R. (1994) *Manufactured Pleasures: Psychological Responses to Design* (Manchester: Manchester University Press).

Dewey, J. (1958) *Art as Experience* (New York: Capricorn Books).

Frascara, J. (2000) "Information design and cultural difference." *Information Design Journal*, Milton Keynes, 9 (2–3), 119–27.

Hermanuz, C. (1996) "Housing for a postmodern world." In Agrest, D., Conway, P. and Kanes Weisman, L. (eds) *The Sex of Architecture* (New York: Abrams).

Illouz, E. (1997) *Consuming the Romantic Utopia: Love and the Cultural Contradictions of Capitalism* (Berkeley, Calif.: University of California Press).

Kaplan, R. (1989) "The analysis of perception via preference: a strategy for studying how the environment is experienced." In *Landscape Planning*, 12(1), 174–5.

Langer, P. (1984) "Sociology—four images of organized diversity: bazaar, jungle, organism, and machine." In Rodwin, L. and Hollister, R. M. (eds) *Cities of the Mind: Images and Themes of the City in the Social Sciences* (New York: Plenum Press), 97–118.

Ley, D. (1981) "Behavioral geography and the philosophies of meaning. Behavioral problems." In Cox, G.R. and Golledge, R.G. (eds) *Geography Revisited* (London: Methuen), 209–30.

Lynch, K. (1981) *Good City Form* (Cambridge, Mass.: MIT Press).

Massey, D. (1994) *Space, Place and Gender* (Cambridge: Polity Press).

McLeod, M. (1996) "'Other' spaces and 'others.'" In Agrest, D., Conway, P. and Kanes Weisman, L. (eds) *The Sex of Architecture* (New York: Abrams).

Mead, G. H. (1934) *Mind, Self, and Society: From the Standpoint of a Social Behaviorist* (Chicago: University of Chicago Press).

Merleau-Ponty, M. (1962) *Phenomenology of Perception* (London: Routledge & Kegan Paul).

Neisser, Ulrich (1976) *Cognition and Reality: Principles and Implications of Cognitive Psychology* (San Francisco: Freeman).

Ockman, J. (1996) "Mirror images: technology, consumption, and the representation of gender in American architecture since World War II." In Agrest, D., Conway, P. and Kanes Weisman, L. (eds) *The Sex of Architecture* (New York: Abrams).

Park, R. E., Burgess, E. W. and McKenzie, R. D. (1996) *The City* (Chicago: University of Chicago Press).

Pile, S. (1996) *The Body and the City: Psychoanalysis, Space and Subjectivity* (London: Routledge).

Rose, G. (1993) *Feminism and Geography: The Limits of Geographical Knowledge* (Minneapolis: University of Minnesota Press).

Sassen, S. (1996) "Analytic borderlands: race, gender and representation in the new city." In King, A. D. (ed.) *Representing the City: Ethnicity, Capital and Culture in the 21st-Century Metropolis* (London: Macmillan Press), 183–202.

Scott Brown, D. (1996) "Through the looking glass." In Agrest, D., Conway, P. and Kanes Weisman, L. (eds) *The Sex of Architecture* (New York: Abrams).

Simmel, G. (1955) *Conflict and the Web of Group Affiliations* (New York: Free Press).

Thrift, Nigel (1996) "On the determination of social action in space and time." *Environment and Planning D: Society and Space*, 1(1), 241–50.

Torre, S. (1996) "Claiming the public space: the mothers of Plaza de Mayo." In: Agrest, D., Conway, P. and Kanes Weisman, L. (eds) *The Sex of Architecture* (New York: Abrams)

Valentine, G. (1992) "Images of danger: women's sources of information about the spatial distribution of male violence." *Area*, 24, 22–9.

Williams, S. J. (1998) "Emotions, cyberspace and the "virtual" body: a critical appraisal." In Bendelow, G. and Williams, S. J. (eds) *Emotions in Social Life: Critical Themes and Contemporary Issues* (London: Virago).

Wilson, E. (1991) *The Sphinx in the City: Urban Life, the Control of Disorder, and Women* (London: Virago).

# 9 Wayfinding research and design

## An interdisciplinary approach in the development of design knowledge and its application

*Romedi Passini*

### Introduction

An interdisciplinary approach in the design field does not consist of just having a person from the social sciences lecturing to designers. This was tried in architectural schools during the late 1960s with mediocre results at best. Knowledge gained by research has to be made accessible and relevant to design, and even more important, it requires that the right questions are posed in the first place. There is a research side to an interdisciplinary approach and there is also a design side. This was well understood in the early 1970s by the emerging discipline of Environmental Design and Research, as it can be seen in the yearly proceedings of the Environmental Design and Research Association since 1970.

In this chapter I would like to emphasize the importance of a collaboration between (1) the research fields, (2) the research fields and design and (3) the design fields. This will be done by referring to my research and design topic: wayfinding and spatial orientation.

### Wayfinding and spatial orientation: an important design issue

It is only in recent years that the extent of wayfinding problems and disorientation, their nuisance and also their functional and financial costs are starting to be recognized. Most people find that wayfinding difficulties and disorientation are highly stressful even in benign cases when the user of a setting is merely confused or delayed. Total disorientation and the sensation of being lost can be a frightening experience and can lead to quite severe emotional reactions including anxiety and insecurity. Self-esteem and assessments of competence may also be affected.

Given the emotional dimension in wayfinding, it is not surprising to find that people include wayfinding criteria when assessing and judging the general quality of a setting. Many organizations catering to the public have become aware that a poor judgement of wayfinding in the setting affects the way the organization itself is perceived. It is quite easy to find publicity advertising "easy wayfinding" in hospital settings, commercial malls, museums and other large public settings to convince the user that the organization is user-friendly. Wayfinding qualities are furthermore associated with levels of accessibility of settings. Many people avoid settings in which they have had experiences of getting lost. The problem of accessibility is emphasized for people with physical or sensory dysfunction.

## Wayfinding and spatial orientation: a historical sketch of the concept from a research perspective

In the general scientific literature the notion of wayfinding was preceded by spatial orientation. Spatial orientation refers to a person's ability of mentally imagining and representing a physical setting and of situating him or herself spatially within that representation. The psychological concept for this mental representation is the cognitive map (Downs and Stea 1977).

The very first references to spatial orientation without a specific reference to cognitive maps date from over a century ago and were written by neurologists who reported cases of patients who, as victims of brain lesions, were incapable of even the most elementary understanding of where they were. Among these authors are: Foerster (1890), Meyer (1900) and Holmes (1918). Case studies of particular lesions have continued in neuropsychology up to the present.

Specific reference to a cognitive map in spatial orientation came from psychology. They were of a descriptive nature at first, without empirical support: Trowbridge (1913), and Griffin (1948). It is Tolman (1948) who in a classic experiment with rats demonstrated the importance of the cognitive map concept.

The next major contribution to spatial orientation came from Urban Planning. Planners preferred the term image as a substitute to cognitive maps. Its major representative was Kevin Lynch (1960) who in a series of cognitive mapping studies of American cities identified the physical elements people use in mentally constructing the image of their city. The five elements he identified (landmarks, paths, limits, nodes and districts) were assumed to be the components of a "legible" and "imageable" city and are still referred to in urban design.

Many studies on cognitive maps were developed by planners and psychologists until the mid-1970s and early 1980s when a certain paradigm shift occurred in the study of spatial orientation and related cognitive research. This shift was brought about by a methodological critique of studies on cognitive maps, in particular the difficulty of measuring spatial representation and of discerning the impact of the mode of expression, such as sketching, describing, or modeling (Moore 1979; Evans 1980). The second argument came from the anthropological literature. It showed that people getting around in monotonous environments like the snow fields of the north, the desert and particularly the ocean (navigation in the South-Seas) could not operate on the basis of a spatial representation (cognitive map) of an undifferentiated environment but had to rely on knowing what to do in order to reach destinations (Gladwin 1970; Lewis 1975).

The convergence of these arguments led to the notion of wayfinding which emphasized the processes involved in reaching destinations. Wayfinding, we have argued, can be modeled as being composed of three major processes:

1 decision-making and the development of a plan of action;
2 decision execution transforming decisions and the decision plan into physical behavior; and
3 information-processing, comprising environmental perception and cognition, underlying both decision-related processes.

Wayfinding can be seen as a problem-solving process with a particularity: it operates in space and requires spatial information. The processing of information, thus, has to include representations of space, that is, cognitive mapping. The feeling of disorientation and being lost is, thus, the consequence of not having a cognitive map and of not having or not being able to develop a decision plan to get somewhere (Passini 1982).

This short sketch of the evolution of spatial orientation and wayfinding illustrates how different disciplines have contributed to the development of the concept. Had we left it to the neurologists, we would have known about the brain, had we left it to the psychologists we would have known about basic cognitive processes, had we left it to the anthropologists, we would have known about the customs of the inhabitant of the South Seas, had we left it to the graphic designers, we would have known about typography.

This is no doubt a caricature but I would argue that the evolution could not have happened without people who had both design experience and research experience. People who were able to formulate problems and develop concepts that were theoretically sound, could be researched, and were relevant to design.

## Wayfinding from a design perspective

Wayfinding in buildings and cities never was a priority in the eyes of architects and urban designers. Wayfinding tended to be associated with signage, and even there the introduction of signage often was, and still is, a last minute thought; a necessary evil, to be watched, so as not to "disfigure" a building. People who had difficulties getting around were seen as being deprived of a sense of orientation and just had to blame themselves. Attitudes are changing and wayfinding is becoming a major design issue.

The concept of wayfinding as defined above points to two major design aspects: (1) the organization of the space and the circulation system created by architects and planners defining the problem people have to solve, and

(2) environmental communication designed by architects, graphic designers, information designers and others providing the information for people to solve the problem.

The conceptualization of the spatial organization and the circulation system appear early in the design process. It is already at this point that the wayfinding tasks of the future users are determined.

## Signs and graphic support-systems

If wayfinding is defined as a spatial problem-solving process, graphic support-systems combined with architecture have to provide the information to solve wayfinding problems, that is to make decisions and to develop a decision plan for unfamiliar routes and to execute decisions on familiar routes. In this respect, three major questions must be answered when conceiving a graphic support-system:

1 what information must be provided;
2 where should the information be; and
3 under which form should it be presented.

The "what" question refers to the content of the message, the "where" question refers to the location of that message. Both are questions pertaining to decision-making and decision execution in wayfinding. The "form" question concerns the actual design of the message in signs, maps, directories, or super graphics, and can be related to knowledge in environmental perception and cognition.

We will just look at the content and location requirement of information. Much of the form question is discussed in *Wayfinding, People, Signs and Architecture* by Arthur and Passini (1992).

## The content and location of wayfinding messages

Information is needed to make wayfinding decisions and it is needed at the moment (location) the decision is made. Decisions are not made in isolation. They are linked together and it is through their links that they become meaningful in a problem-solving situation. Taking an elevator, for example, illustrates clearly that a series of decisions executed in a prescribed order are necessary to complete the task, that is, "pressing a call-button," "entering the cabin," "pressing the destination floor number on the control board," "checking the attained level," "getting out at the desired level." This order represents a kind of blue-print for the wayfinding task: taking an elevator. In our jargon we call this blue-print a decision plan.

Taking an elevator, however, is probably not a main wayfinding task. It may be related to other decisions such as: "to enter a building," "to find the location (address) of a destination," and once the indicated floor is reached by taking the elevator, it might be related to the decision "to follow a corridor to the indicated door number." These decisions form a blueprint or decision plan to complete the original task of reaching a given destination.

If one respects the logic between decisions, that is, if one links decisions according to an "in-order relation" (in order to do "A", I do "B") one finds that wayfinding decisions are hierarchically structured. At the top is the original wayfinding task, at the bottom of the structure are the decisions leading directly to behavioral actions. In between are the "higher order decisions" which are sub-tasks when viewed from the bottom, or part of a decision plan if viewed from the top.

In fact, hierarchical structures of decisions underlay most purposeful cognitive endeavors. They decompose complex problems into manageable sub-problems; inversely, they allow for treatment of a sub-problem while keeping in mind the overall solution. Hierarchical structures are also great mnemonic devices. They help remembering decisions and information, which is particularly important when learning new routes. A decision plan indicates how the person has solved a wayfinding problem. If we accept that the purpose of graphic, and environmental communication in general, is to provide the information for decision-making in wayfinding, we must conclude that the content of the messages on information displays should correspond to a person's wayfinding decisions.

Following this line of reasoning, we can state that a graphic support-system is an ensemble of information displays corresponding to a typical decision plan. The logic that links the decisions into a plan is the same logic that links information displays into a system.

If the content of required messages in graphic supports is given by the wayfinding decisions, it is those points along a route which require a decision that determines the location of the message. We have ample evidence that signs are not seen just because they are there, they tend to be seen when they are needed. Environmental perception has to deal with more information than can be consciously registered, this is especially true in complex indoor environments. People are not just exposed to the environment like a camera, rather they select and choose. One of the major selection criteria is the usefulness of the information at a particular moment in time. The optimal location of information displays, therefore, is again determined by the decision plan.

I hope that I have been able to illustrate how an interdisciplinary concept not only provides information to designers but can lead to a new approach to design.

## Interdisciplinarity in the design profession

Throughout this chapter we have argued that if the spatial organization of a setting and the circulation system define the wayfinding problem the user will have to solve, environmental communication provides the information the user will need to solve the problem.

This description of wayfinding design shows the inevitable link between the two interventions. Environmental communication or information design cannot be seen independently of spatial organization and the circulation system. Environmental communication, furthermore, concerns not only graphic design but also architecture.

Design for wayfinding is part of the architectural profession including urban planning and landscape architecture as well as graphic design, and it is time for a collaborative effort among the design disciplines. The practice of letting graphic designers install a few signs hours before opening day should be vigorously denounced. Both design disciplines should know about each other's functions and procedures and both should know about their contribution to wayfinding design.

Wayfinding has been chosen as an illustration; I am sure, the same argument could be made for many other concepts involving the social sciences and design.

## References

Arthur, P., and Passini, R. (1992) *Wayfinding: People, Signs and Architecture* (Toronto: McGraw Hill).

Downs, R., and Stea, D. (1977) *Maps in Mind* (New York: Harper and Row).

Evans, G. W. (1980) "Environmental cognition." *Psychological Bulletin*, 88(2), 259–87.

Foerster, R. (1890) "Über Rindenblindheit." *Archiv für Ophthalmologie*, 36, 94–108

Gladwin, T. (1970) *East is a Big Bird: Navigation and Logic on Puluwat Atoll* (Cambridge, Mass: Harvard University Press).

Griffin, P. (1948) "Topological orientation." In E.G. Boring *et al.* (eds) *Foundation of Psychology* (New York: John Wiley and Sons), 380–6.

Holmes, G. (1918) "Disturbances of visual orientation." *British Journal of Oftalmology*, 2, 449–516.

Lewis, D. (1975) *We the Navigators: The Ancient Art of Landfinding in the Pacific* (Honolulu: The University Press).

Lynch, K. (1960) *The Image of the City* (Cambridge, Mass.: MIT Press).

Meyer, O. (1900) "Ein- und doppelseitige homonyme Hemianopsie mit Orientierungsstörungen." *Monatsschrift der Psychiatrischen Neurologie*, 8, 440–56.

Moore, G. T. (1979) "Knowing about environmental knowing: the current state of theory and research on environmental cognition." *Environment and Behavior*, 11(1), 33–70.

Passini, R. (1982) *Wayfinding in Architecture* (New York: Van Nostrand Reinhold).

Tolman, E. C. (1948) "Cognitive maps in rats and men." *Psychological Review*, 55(4), 189–208.

Trowbridge, C. C. (1913) "On fundamental methods of orientation and imaginary maps." *Science*, 38(990), 888–97.

# 10 Preventing drug interactions in older adults

## A collaborative study[1]

*Zoe Strickler and Patricia Neafsey*

> Most psychological theories were cast long before the advent of enormous advances in the technology of communication. As a result, they give insufficient attention to the increasingly powerful role that the symbolic environment plays in present-day human lives.
>
> (Bandura 1986: 20)

## Introduction

The study Preventing Drug Interactions in Older Adults, is an interdisciplinary project that demonstrates the potential for collaboration between social scientists and designers on experimental communication research. This chapter describes the nature of the collaboration and discusses preliminary findings regarding design for interactive learning materials for older adults. The central question of the study is whether an animated, interactive software program can be effective in helping persons 60 and older to prevent common, harmful drug interactions among prescription medicines, over-the-counter (OTC) medications, and alcohol. The project is funded by the Donaghue Medical Research Foundation.

This study originated six years ago when co-author Patricia Neafsey, a pharmacologist in the School of Nursing at the University of Connecticut, reflected on her experiences teaching drug interaction information to students in the undergraduate nursing program. It is important to note, and not unusual, that a project that is, to a great extent, a communication design study, originated outside the field of communication design. The design academic community has been slow to respond to the growth of communication research (Winkler 1997) as it has emerged as a component of studies across the spectrum of the social and health sciences disciplines (Strickler 1998). However, as an increasing number of studies in fields from social psychology to epidemiology require communication components, the design academic community should begin to prepare faculty and graduate students to contribute to this research in meaningful and formative ways. This chapter discusses an experimental collaboration between a pharmacologist, a visual communication designer, and an educational psychology team on an intervention to prevent drug interactions in independently living older adults.[2]

## Origins of the project

The problem that sparked the present study was a pedagogical one. Students in the clinical science courses at the University of Connecticut consistently had difficulty mastering material on adverse effects among prescription and (OTC) drugs, alcohol, and certain nutrients. Drug interaction information can be perplexing when approached as a memorization problem because of the large number of substances involved and the subtle differences among possible conflicts. Drugs with differing pharmacological effects may be used to treat the same symptom, whereas substances that are chemically different can sometimes work in the same way. Confusion between scientific names and brand names confounds the problem for the typical learner.

Dr. Neafsey began to look at the sources of the students' difficulty and to consider alternative ways to present course material. She worked with a local high school computer class to create simple animations of the pharmacological effects of several groups of pharmaceuticals implicated in common, but serious interactions. The animations provided students with visually-reinforced conceptual frameworks for understanding the mechanisms of these basic classes of substances. When the animations were embedded in an interactive software program that could be used along with the traditional lecture format nursing students' comprehension of the material improved noticeably.

Neafsey and Robin Froman, an educational psychologist and statistical consultant, conducted a longitudinal, experimental study to assess knowledge and self-efficacy outcomes (confidence in one's ability to apply the information) among undergraduate RN students receiving only traditional lectures versus those using the software program along with lectures (Froman *et al.* 1993; Neafsey 1997; Neafsey 1998). The computer-aided learners showed greater gains and longer retention than those receiving only traditional instruction.

Neafsey then began to ask whether a similar program could reduce the actual incidence of risk behaviors in a population of patients. It seemed that if nursing students at a research university had difficulty learning relations among medicines, then the average person would have even more difficulty. Likewise, if a multimedia software program could improve performance among college students, such a program might potentially be a useful behavioral intervention with high-risk patients.

## The scope of the problem

Thus began the present study of whether an interactive software program designed for persons 60 years and older might constructively change dangerous self-medication behaviors in that population.

The study population was selected because older persons are particularly vulnerable to injury from interactions among pharmaceuticals. This is because they are more likely to use multiple prescription medications for age-related chronic conditions such as high blood pressure, and their drug metabolism rates are more variable than for members of the general population. They are often less able to hear, read, and understand oral and written instructions (Bloom *et al.* 1993; Hanlon *et al.* 1992; Pollow *et al.* 1994; Salzman 1995; Wallsten *et al.* 1995).

Other problems arise for seniors from interactions with alcohol and OTC agents. Studies of alcohol use among older adults indicate that approximately 65 percent self-report use of alcohol on a regular basis (Forster *et al.* 1993; Lisansky Gomberg 1995; Moore *et al.* 1999). Adams (1995) found 38 percent of 311 residents of retirement communities who responded to a mail survey reported using alcohol and a "high risk" medication likely to interact with alcohol. Although national surveys of older Americans living independently suggest that nearly 75 percent of those aged 60 and older consider their health to be good to excellent, and 90 percent feel they are doing a good to excellent job of taking care of their health (National Center for Health Statistics 1993), serious interactions between prescription medications with other OTC agents and alcohol are common in this age group (Manasse 1995; Taskforce for Compliance 1994; USGA 1995).

A further problem arises in that many medications that were previously available only by prescription are now approved for OTC sales. Exposure to television and magazine advertisements for over-the-counter agents has increased their use by older persons, although dosage recommendations appropriate for the general population may be inappropriate for persons with slower body metabolism and/or complex prescription medication regimes.

Some interaction information is available on most drug packaging and advertisements. However, studies of functional health literacy of older adults suggest that the majority of older consumers do not understand what they read on medication packets and inserts (Gazmararian *et al.* 1999; Jackson *et al.* 1994; Kirsch *et al.* 1993; Williams *et al.* 1995). In general, older adults were found to comprehend health information a full four years below their reading grade level of non-technical prose (Williams *et al.* 1995). Considering the frequency of functional health illiteracy among older adults, researchers recommend that health information be presented at no more than a sixth grade reading level (Laubach and Koschnick 1997; Plimpton and Root 1994; Williams *et al.* 1995), yet, currently, package inserts and label warnings are written only in English and at a 12th or higher grade level (Davis *et al.* 1990; Gazmararian *et al.* 1999). In addition to problems of comprehension, the type sizes on these print materials are often so small they cannot be read by persons with failing eyesight.

Adverse drug interactions have significant costs for older adults and for society. Adverse drug reactions account for 17 percent of hospital admissions for the elderly, almost six times more than for the general population (USGA 1995). Failing to take medications properly is estimated to cost the health care system $25 billion annually, and results in 10 percent of nursing home admissions costing $5 billion a year (Taskforce for Compliance 1994). Drug interactions rank between the fourth to sixth leading cause of death in persons 65 and older (Lazarou *et al.* 1998). Thus, the self-report by older persons of good health and ability to provide safe self-care do not reflect the reality of existing problems of drug management for this segment of the population.

## The role of design

Since the goal of the research was to assess the effectiveness of interactive software as a learning tool for active seniors and as an instrument of health behavioral change, the matter of the suitability of the software for older users was central to the study. It was determined early on that a communication designer would need to be involved in the study and that the program should be thoroughly pre-tested in order to avoid basic research errors (i.e. that the phenomena being measured or the instruments used to measure them would be found inadequate to answer the research question). In other words, the validity of this study would rest, in part, on the design of a program that was suitable for older users. It is important to note that the interactive, animated program created for the clinical science courses effected change in student performance, although it was not designed by professional designers. The question of whether, and to what extent, visual qualities of a communication vehicle influence behaviors and performances is a largely unmapped territory in experimental research. These factors are unlikely to be explored until designers act as generative players in research development. However, in addition to measuring responses to designed products in such studies, design researchers must also measure responses to "vernacular" and "naïve" variations of the same communication, in order to determine the extent to which design factors influence behavior.

Environmental considerations are important here too. Students working toward a grade in a course are a captive and motivated audience. In this environment naive design work may be effective in ways that it would not in environments in which users choose whether to pay attention to a communication, and make their evaluations of it relative to prior experiences with commercial products and communications.

A review of existing literature revealed that while a body of research exists on aspects of computer use by older adults with respect to various age-related perceptual disabilities, few studies have been conducted to address the aesthetic and visual attributes of interactive programs that make them attractive to and effective for older users. Questions that are generally not addressed include: What should the software program look and feel like for the user? What specific aesthetic properties contribute to a user's "liking" of the program? What visual features enhance the users' attention to the information and improve ease of use, comprehension, and retention of the information content? Answers to these types of questions would aid designers working to incorporate research findings into learning materials for aged users.

In the studies reviewed, the predominant approach to studying older persons' use of computers was to measure specific stimulus responses or task performances in isolation. While focused investigations of this sort, conducted in laboratory environments, yield findings that are reliable and reproducible, they leave open questions of generalizability and realism. In other words, will the results hold in natural settings where the phenomena are embedded in real world contexts and products? Do the findings provide valid guidelines for designers?

Designers have not historically contributed to experimental social research in instrumental ways—certainly not as full collaborators. Therefore, a practical hole remains in much literature concerning the effects of communication media on learning and behavioral outcomes. Furthermore, what literature of this sort exists has not been effectively organized and cataloged for easy access by the design profession (Poggenpohl 1998). The literature gap with respect to design for older adults is addressed throughout *The Handbook of Human Factors and the Older Adult* (Fisk and Rogers 1997). In Chapter 2 Gardner-Bonneau and Gosbee write: "Applied literature is particularly sparse in this area, and it is our view that significant applied research efforts are needed to understand the health care and rehabilitation systems existing in the United States, so that the human factors issues in those systems, as they apply to the nation's elderly, can be better addressed . . . including medical communication systems" (253).

They further discuss the issue of transference between researchers and designers: "There is a definite need to learn more about the performance capabilities and limitations of older adults in real-world settings as opposed to the contrived situations of the laboratory. . . . For your research to have an impact on the lives and performance of the elderly in health care and rehabilitation settings, your results must be quantifiable and communicated to designers in such a way that they can use the information. That is, designers need . . . not only the results, but some indication of the generalizability of the results and any situational or human constraints that limit their applicability" (249).

Preventing Drug Interactions in Older Adults is then, in part, an effort to build design variables into a study from its inception in order to evaluate the effects of design decisions on users, first qualitatively in the pre-test phase, and then in context, in an extended clinical trial.

## Interdisciplinary research plan

A defining feature of the project is its interdisciplinary approach. The information outcomes—what we will know at the end of the study—could not be obtained from within any one discipline. A diagram of the experimental design (Figure 10.1) shows the interrelation of the disciplines required. A classical experiment in an older field such as physics might draw theory, methods, and substantive content from within a single discipline: theory from theoretical physics, method from experimental physics, and substantive content from physics, e.g. subatomic particles, forces, etc. In Preventing Drug Interactions in Older Adults, theory, methods, and content are drawn from at least six different disciplines. Although the emergence of subspecialties in the physical sciences such as biochemistry and physical chemistry suggests that historical divisions are breaking down even in the "hardest" of sciences, design research is inherently interdisciplinary because the problem content always comes from another discipline. This sort of disciplinary complexity defines communication research generally.

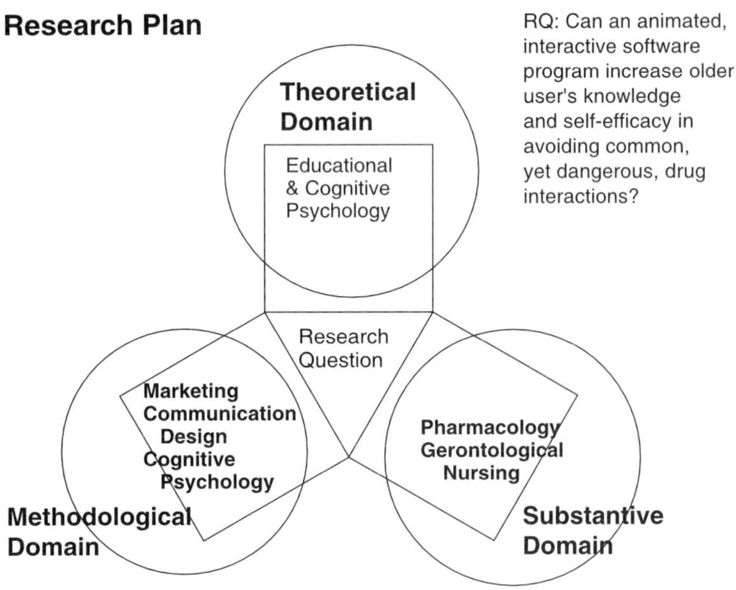


*Figure 10.1* Research plan.

## Hypotheses and goals

Hypotheses for the study were stated as follows:

1 Subjects using the Personal Education Program (PEP) will show greater knowledge of potential interactions of alcohol with prescription and over-the-counter medications than non-users.
2 Users of the Personal Education Program (PEP) will show greater self-efficacy for how to avoid drug and alcohol interactions than non-users.
3 Users of the PEP will self-report fewer behaviors associated with self-medication interactions than non-users.

## Research questions

At the conclusion of the project we will know whether the PEP changes knowledge, self-efficacy, and behaviors in older users, and how the PEP works differentially for persons of various age, gender, race and educational backgrounds. In addition, we will know if a PEP could feasibly and effectively be placed in a community setting and what the barriers to use are. Finally, we will know what changes could/should be made to the program delivery to increase effectiveness.

## Theory

An organizing theory for the study is Bandura's conception of self-efficacy as it applies to learning and task performance. Self-efficacy, or the perception of one's ability to successfully complete a task, is a construct central to Social Cognitive Theory which considers the influence of self-reflective processes on behaviors (Bandura 1986, 1997). Self-efficacy is related to whether an individual attempts tasks, persists, and achieves successful completion (Bandura 1977a, b, 1986, 1996 *et al.*, 1997; O'Leary 1985: Strecher *et al.* 1986). Self-efficacy has been shown repeatedly to offer explanation of performance of behaviors beyond knowledge alone (Strecher *et al.* 1986). Such findings and others have been repeatedly interpreted to mean that both knowledge and self-efficacy are important and different cognitive precursors of behavior (Bandura 1986, 1996 *et al.*, 1997; Froman 1996; Murdock and Neafsey 1995; Neafsey 1997; Strecher *et al.* 1986). Bandura's theory of self-efficacy can be summarized with the following statement: knowledge + increased likelihood of successful completion of behavior = self-efficacy.

Self-efficacy is not a fixed personality trait as is self-esteem (the perception of self-worth). Whereas a person's base level of self-esteem remains relatively constant across situations, self-efficacy is sensitive to particular social environments and tasks. People can believe themselves to be efficacious (competent to succeed) in one task or skill area, but inefficacious (incompetent) in another (Bandura 1986, 1997). Although base levels of self-esteem have been found to correlate to a small degree with self-efficacy, self-esteem is generally a poor predictor of self-efficacy as it relates to particular tasks and skill sets. In this context, the matter of whether a task, particularly a new task to the individual, seems relatively easy to master affects whether a person will believe that he or she can complete the task well enough to actually try it, and especially whether he or she will persevere with it (Bandura 1986, 1997).

For designers, the theorized role of self-efficacy in these processes should be regarded as a call to action. Consideration of people's emotional and aesthetic responses to designed objects or communications is the traditional purview of the designer, and, by extension, this concern should include psychological responses that influence behaviors. Yet, designers have played little part in experimental research efforts to date in subject areas concerning beliefs, attitudes and behavioral change. The significant point here for designers is that the need for effective design is embedded in the organizing theories behind much contemporary behavior research (Strickler 1998). Designers should be increasingly involved in this work to ensure that the formal and functional variables of communications employed in the studies are controlled for as thoroughly, and are as valid, as survey and interview instruments used for measurements.

## Study design

Preventing Drug Interactions in Older Adults has two distinct phases: Phase 1, a pilot study, and Phase 2, an extended clinical trial. During Phase 1, formative research was conducted for design of the PEP and development of measurement instruments. A pilot study of one segment of the PEP was conducted with 60 subjects in the fall of 1998. Phase 2 consists of an extended three-year clinical trial of the revised, full PEP. The study design can briefly be described as follows.

### *Pilot year: development of quantitative measures*

During Phase 1 of the study, three instruments to measure outcomes were developed and tested for use. These were: (a) an objective test to measure subjects' knowledge of drug interaction content; (b) a self-report measure of self-efficacy; (c) a self-report measure of prescription and OTC drug use and alcohol use to study changes in patterns of self-medication; and (d) a measure of user satisfaction about details of the PEP.

Additionally, a survey of computer use, attitudes, and access by older users was developed to help understand older adults' perceptions of benefits and barriers to use a computer equipped with a touch screen. Design and testing of these instruments are described in detail elsewhere (Neafsey *et al.* in press).

*Development of PEP through qualitative pre-study*

Also during Phase 1, formative focus group research was conducted with representative older participants to guide design of the PEP in three areas. Areas investigated were: (a) participant's existing levels of knowledge regarding drug interaction information; (b) participant's language comprehension levels; and (c) participant's aesthetic and functional preferences for design of the computer interface.

Methods and findings for the qualitative pre-study are discussed briefly in this paper and, in detail, elsewhere (Strickler and Neafsey 2002).

### *Extended clinical trial: measuring knowledge and behavioral outcomes*

In Phase 2, the extended clinical trial currently under way will gather information on the effectiveness of the PEP in two different types of settings: in patients' homes and in a health clinic environment. Both the home study and the clinic study will be conducted through the Visiting Nurse Association (VNA) of Connecticut. Subjects for both segments are to be recruited in three separate communities within the state of Connecticut—Enfield, Willimantic, and New Britain—to ensure recruitment of older adults with varying educational, ethnic and racial characteristics. In the home study University of Connecticut seniors and graduate nursing students carry laptop computers equipped with touch screens into older adults homes. In this assisted learning environment, participants are able to ask questions about their particular pharmaceutical regimens while using the program. In the clinic study patients at VNA blood pressure clinics will be able to use the program individually for self-directed learning. The computers will be equipped with a tracking program that will keep a record of each user's path through the program. Individuals tend to return to the same blood pressure clinic which makes long-term subject evaluation possible.

The study is a repeated measures design with two between factors (site, group) and one within factor (time). Subjects are evaluated immediately after treatment (immediate post-test) and again one and two months later (delayed post-tests).

*Three test groups*

In both studies, home and clinic, participants will be randomly assigned to one of three test groups: (1) an experimental group that will use the PEP along with a printout of the information, or (2) and (3) either of two control groups that will not use the PEP. One control group will receive the drug interaction information contained in the PEP in print form only. The other control group will receive no intervention during the course of the study. All three test groups will receive an identical series of knowledge and self-efficacy instruments with the same time intervals between (1 month). The number of participants in each test group will be 33. This allows for a 33 percent attrition rate for the study. A final cell size of 22 provides sufficient power (0.80) to detect medium to large differences between groups.

A primary goal of the study is to compare information retention and reported behavioral change between the experimental group using the computer medium (in combination with print) as compared to those using print only. It is assumed that both of these groups will perform better than the group receiving no intervention. At the end of the study, members of the two control groups will be given the opportunity to use the PEP for their personal benefit.

### *Study population*

The population for whom the program is intended is, by broad definition, the independently living older adult managing his or her own medication regimen. With the understanding that one software program cannot address extremes of need and performance capability within the older community, outer parameters of need and competency were established to define the user population.

Subjects selected for the study are at least 60 years of age by self-report. Criteria developed and validated by the MacArthur Research Program on Successful Aging are being used to identify older adults with independent physical and cognitive functioning (Wallsten *et al.* 1995). Study participants are able to: (1) perform activities of daily living on the Katz tool (1970), and seven of eight combined Nagi and Breslau functional items (Nagi 1976; Rosow and Breslau 1966); (2) answer six of ten items on the Short Portable Mental Status Questionnaire (Pfeiffer 1974); (3) have a reading comprehension level of at least grade 6 on the Rapid Estimate of Adult Literacy in Medicine (REALM) (Davis *et al.* 1993); and (4) be living independently. Participants are told that the REALM is to evaluate what health care information is clear or confusing so that we may write the PEP more clearly. The REALM was chosen because of its documented reliability and validity for rapid assessment of reading comprehension (Davis *et al.* 1990; Davis *et al.* 1993). Subjects are screened for visual acuity using a pocket vision screener (Rosenbaum, Graham-Field Surgical Co., New Hyde Park, NY). To be included in the study, subjects must have a visual acuity of 20/100, with corrective lenses as needed.

Study participants also need to meet the following criteria: (1) take a prescription antihypertensive or anticoagulant, and (2) take analgesics (aspirin-like pain relievers such as ibuprofen, or non-aspirin pain relievers such as acetaminophen), antiacids, or acid reducers. These additional criteria are invoked as they reflect the audience who would most benefit from the program and be interested in, engaged by, and motivated to use the content. The PEP vehicle can later be adapted for any variety of content areas relevant to community living older adults (diet, exercise, pain management, etc.).

It is assumed that the population served by this PEP will be individuals proficient in the written English language and of a socio-economic status high enough to be in the US health care system (i.e. taking prescription medications). It is understood that a PEP meeting quite different design criteria would be necessary to deliver health information to non-english speaking persons and to those with less income or education.

## Utility of qualitative and quantitative data for design purposes

For practicing designers, qualitative, descriptive findings from a study are often the most useful for applying knowledge to the design of products. This need, however, runs counter to the way findings are generally reported in scientific studies. Because of the centrality of statistical analysis to establishing rigor in research, many studies do not fully report qualitative findings even when formative pre-studies have been conducted. From the standpoint of a designer, this practice represents a significant loss of data.

Qualitative findings—generally gathered through individual or group interviews with representative members of a study population—provide insight into subjects' thought processes, and emotional responses to phenomena. These findings are typically gathered prior to development of quantitative measurement instruments (e.g. surveys) to ensure validity in the instruments (Fowler 1984; Morgan 1988). Statistical analysis of quantitative measures remains essential in order to answer research questions with scientific confidence. However, knowing that a particular communication design effected change in a population is not as valuable for a designer as knowing which features of the design contributed to the effectiveness and by what mechanisms they were theorized to be effective. Information of this kind might only be gathered through testing of multiple prototypes with alternative sets of visual variables under controlled circumstances. This, however, is an area of communication design research in which methodology has yet to be fully developed.

Measuring alternative design variables is not a procedure other disciplines are likely to approach comprehensively, even though they may use communication elements in their studies. If thorough investigation of design variables is not regarded as necessary to address the central research question in a study, such investigation will not be conducted. Researchers from other disciplines are also less likely to be sensitive to the range of possible visual or aesthetic variables in designed communications or to their significance for user interest, involvement with, and comprehension of information. Certainly, they do not possess the expertise to generate necessary prototypes for representing multiple design variables. These are the values and expertise that designers can bring to communication research that have the potential to make research findings both more relevant to subjects living in a highly visual and media saturated culture, and also more applicable to products produced within that culture.

## The qualitative pre-study

### *Focus groups/formative evaluation*

In the formative pre-study for Preventing Drug Interactions in Older Adults we presented alternative prototypes for visual style and functional variables to focus groups of representative users for evaluation. Two focus groups of six volunteers each (three men and nine women) met once a week for nine weeks to evaluate PEP components as they were being designed and written. Focus group participants were recruited from the Center for Learning in Retirement (CLIR) which is an ongoing program of lectures, seminars and workshops at the University of Connecticut and is attended by over 300 older adults. These individuals may be characterized as active, community living older adults with high motivation to contribute time and participation to research and educational efforts.

The focus groups met in a CLIR seminar room to evaluate micro-elements of the software such as text, font, color, background, complexity of information layers, button styles and locations, feedback to interactions, and animation style, speed, and location. The focus group method of Anderson and colleagues was used (Anderson *et al.* 1996). The focus group participants met all of the study criteria for subjects as listed above. Each of the focus group members was given a $5.00 cash incentive award for each meeting attended. Selected animations from each learning section were revised until group members agreed that the animations were visually appealing, easy to follow and understand, timed correctly for comprehension, and that there were minimal extraneous visual stimuli. The focus groups also evaluated learner prompts and questions embedded in the optional PEP quiz sections, as well as the level and clarity of text used in accompanying printouts.

### *Qualitative findings*

The focus group sessions provided insight into a number of matters affecting participants' use of the computer learning program. Participants spoke openly and matter-of-factly about age-related issues such as diminished visual capability. They communicated their likes and dislikes for graphic style clearly. They also talked about their prior assumptions regarding interactions between prescription and over-the-counter medications and alcohol.

What follows is an overview of some of the design choices and modifications that were made to the PEP in response to the participants' comments and preferences. A thorough discussion of method and findings is reported elsewhere (Strickler and Neafsey 2002).

### *Emotional tone of the design*

Comments made by the seniors in the focus group interviews suggested that a straightforward approach to the content was preferred for this topic. A comment in the first focus group comparing alternative styles of illustration addressed this issue: "At least it looks like a person and not a dog." The animation prototypes presented to the participants were essentially moving anatomical diagrams, although because we used the term "animation" several participants may have expected a children's cartoon approach. Their preference for a straightforward, rather than rhetorical, approach appeared randomly throughout the pre-study in comments such as: "This [animation] doesn't show the heart. I like having the heart. The anatomical one"; or: "That other one, the Tums, was too much like advertisement. We're so used to watching television every day. It's like—take this one—this one is best."

A second category of comment from the participants that had implications for communication tone could be identified in their strong interest in the topic of drug interactions and the high level of salience it appears to have for their daily lives. Statements such as: "Now that I've seen those I want to see the whole thing"; or "I learn something every time I come. I'll never take Tums after I take my [coated] aspirin. Two hours before or two hours after," were typical responses to the sessions. Because of this expressed desire among the participants to learn about the information and to implement new behaviors in their own lives we determined that elements of persuasion would be largely unnecessary, and perhaps off-putting, in the program.

### *Method for arriving at illustration style*

In the first focus group sessions participants were shown three animation prototypes rendered in different illustration styles (see Figures 10.2, 10.3 and 10.4).[2] Figure 10.4 was unanimously liked by the participants, stylistically and functionally. Because of the overwhelmingly positive response to Figure 10.4 we used its basic features for the final program although subsequent revisions were made per participants' recommendations. All features of the animations that evoked comment from the participants were examined and changes were incorporated into the visual style over the course of the nine-week formative period. The final anatomical figure and screen style are shown in Figure 10.5.

A number of the animations used a zoom function to show internal physiological processes. The zoom transition was effected by drawing a circle around the part of the figure to be detailed (e.g. the liver), and then

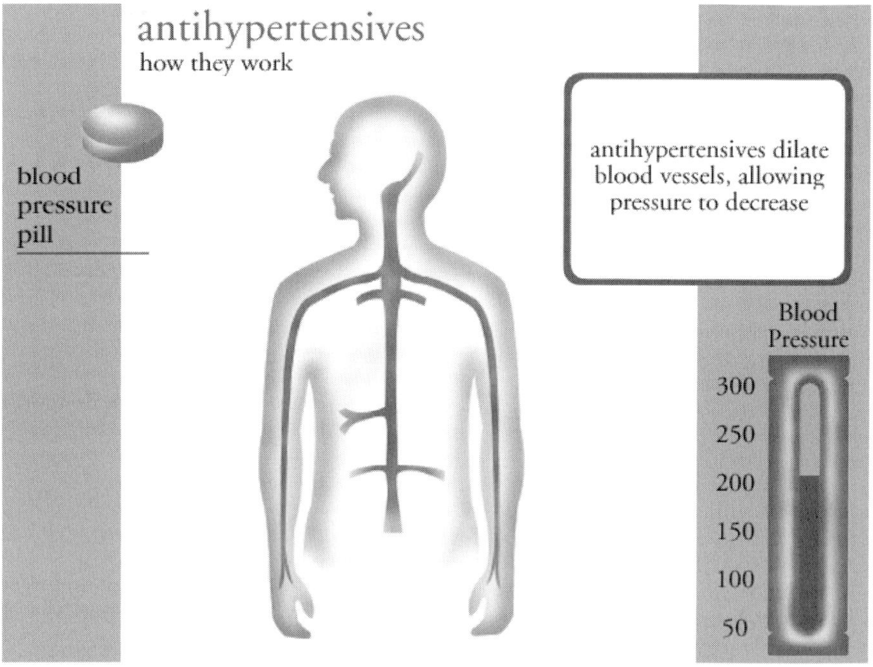


*Figure 10.2* Antihypertensives/1.

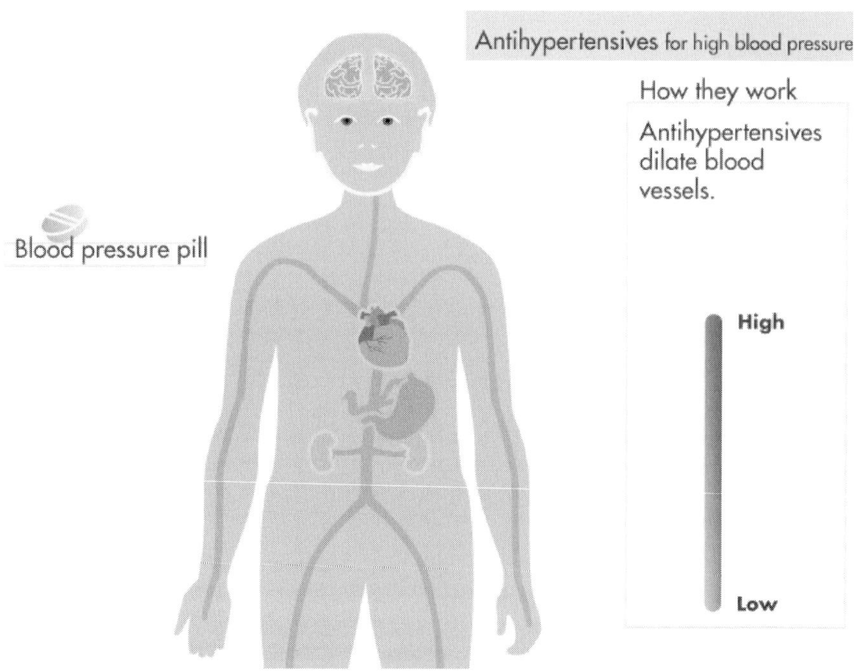


*Figure 10.3* Antihypertensives/2.

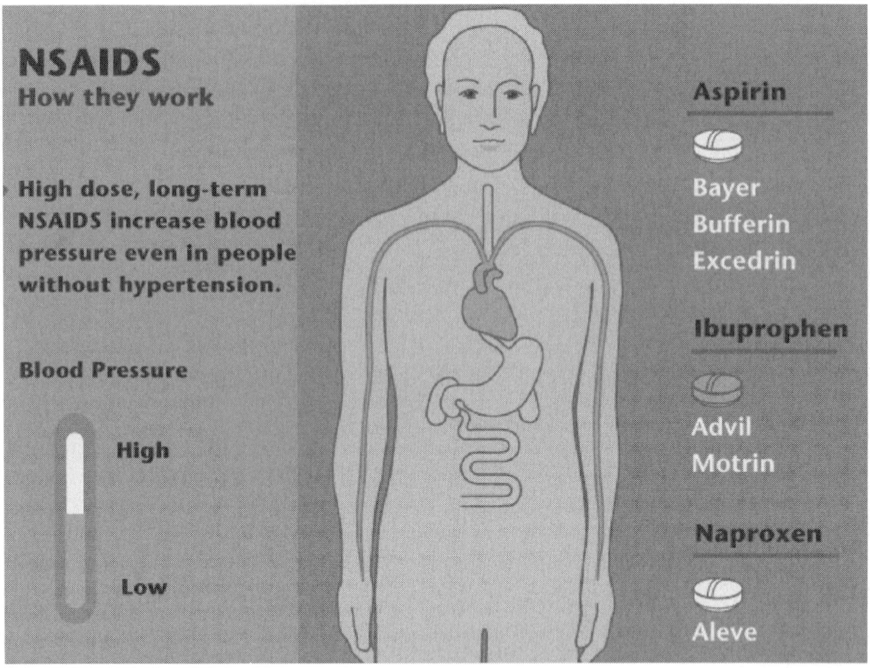


*Figure 10.4* NSAIDS.

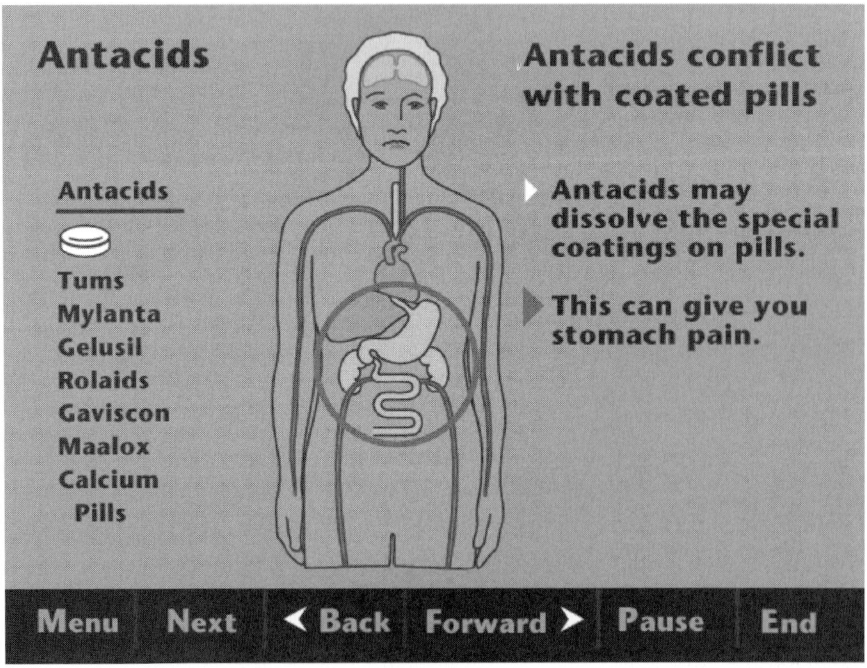


*Figure 10.5* Antacids/1.

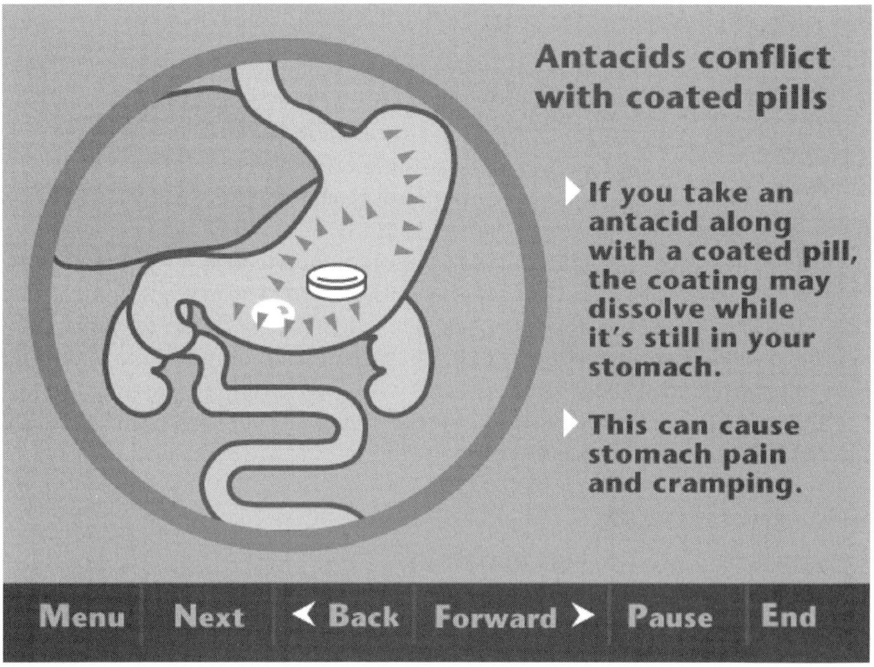


*Figure 10.6* Antacids/2.

zooming the circle while the figure faded. A zoomed detail screen is shown in Figure 10.6. Prototypes for the sessions were designed by Zoe Strickler and graphic design students Michael Skiles and Mai Phung.

## Summary of findings from the focus groups

### *Flat vs. dimensional diagrammatic style*

Participants preferred a flat, planar diagrammatic style to a gradated style that implied dimensionality in the figure. The gradation was regarded as distracting and difficult to see. This is consistent with research suggesting that difficulty distinguishing subtle color variation and low contrast edge discrimination is a prominent visual deficit associated with aging (Kline and Scialfa 1997; Morris 1994).

### *Contrast*

A planar figure with dark, thick outlines around organs and features was preferred over a planar figure in which features were differentiated by color and value alone or by a lighter or thinner outline. This is consistent, as discussed above, with studies suggesting that the ability to perceive subtle contrasts diminishes with age (Kline and Scialfa 1997).

### *Features of the figure*

A figure that appeared to be middle aged was preferred over a figure that had features associated with aging. A figure with distinct, visible hair was preferred over a figure that was bald or had indistinct hair features. A figure that was facing forward with complete facial features was preferred over a figure with the head in profile and absent facial features. A planar figure that was more naturalistic in its body contours and features was preferred over a planar figure that was more geometrically abstract.

It was part of the objective of the design team to create a figure that was largely gender and race neutral. This turned out to be an extremely complex visual problem which deserves further study in its own right. The final figure tested by the focus groups was perceived by participants to be more or less gender neutral, but the issue of race was addressed by alternating figures of different colors in the interactive program.

### *Completeness of figure*

A figure that displayed all the major organs was preferred over a simplified figure that displayed only those anatomical systems discussed in the particular section. Participants expressed specific desire to see the brain at all times so as to know whether the brain was affected by a particular drug interaction. This suggests the understandable concern that older adults have for changes in mental function.

### *Background*

Participants preferred a plain background color of a single hue and value with no layers, shaded bars, or divisions to separate text from elements. Where space was available, participants preferred that type be enlarged for legibility rather than reduced to preserve compositional open space.

Participants strongly disliked prototypes employing a white background as they found the screens glaring. This is consistent with findings that sensitivity to glare and the eye's ability to adjust to intense light sources diminishes with age (Kline and Scialfa 1997; Morrel and Echt 1997; Morris 1994). A warm blue background of medium to light value was preferred over brighter or warmer backgrounds.

### *Typography*

Participants' preferences for type were consistent with existing studies of print samples for older adults (Morrel and Echt 1997). A bold, sans serif font (Stone Sans) was used for text throughout, in sizes no smaller than 18 pt. Body text was 20 pt; heads were set in 32 and 24. The body format was flush left/ragged right, set in small blocks of text for rapid reading. Text lines were typically two to four words in length. The longest block of continuous text was five lines.

Participants expressed a strong interest in seeing particular words in the text emphasized. Because the text was already set bold, increasing weight for emphasis was unpleasant. Participants could not perceive color changes in the text well, a finding consistent with research demonstrating loss of color perception with age (Kline and Scialfa 1997; Morrel and Echt 1997; Morris 1994).

Participants had difficulty perceiving words set oblique and expressed a strong preference for underlining of key words as the primary method of emphasis. This contradicts fine typographic convention, but reflected participants greater familiarity with typewriter conventions, expressed as: "We're used to seeing type underlined for emphasis."

### *Motion*

Because of diminished ability to scan visual fields and to track motion in older adults (Kline and Scialfa 1997) animations were designed so that only one change or event happened at a time, with generous time frames between events (time interval results are reported elsewhere in Strickler and Neafsey 2002). However, they were also sensitive to times when motion or text change appeared too slow. Participants expressed a strong preference and need for visual cueing (in the form of bright red arrows and circles) to guide the eye through sequences of text and animation.

### *Language comprehension level*

Parallel focus groups with the same participants were conducted to test the language level of texts and to assure clarity in verbal explanations of pharmacological and physiological processes. The objective was to keep texts near a sixth grade reading level, and to make verbal descriptions as easy to understand as possible.

### *Touch screen*

Touch screen devices were attached to the IBM laptop computers to eliminate difficulties that older users encounter from keyboard and mouse entry devices (Morris 1994). Description of the technical design and testing of the touch screen attachment is reported in Neafsey *et al.* (2001).

## Implications for designers

Although the design team reviewed existing literature on design of visual learning materials for older adults prior to preparing the prototypes, it was clear from the focus group interviews that simply reading existing literature was insufficient preparation to create successful designs without further testing.

For a number of features of the prototypes, aesthetic values learned in the conventional design classroom proved inappropriate for the application. For example, despite having read studies on loss of visual acuity in older adults, the impulse to make type small was apparent in all initial prototypes produced by the design team. The modernist dictum that type should only be as large as necessary to serve its intended functions (West 1990: 123), leads to an impulse in designers to use type that is often, in reality, too small for the function. When designing for older users within the bounds of a 14-inch computer screen there is an inherent tension between the need to make the type as large as possible for reading and fear of overcrowding the visual area. However, it was clear from the focus group sessions that the older users were substantially less concerned with issues of compositional breathing room than with legibility.

Designers entering into experimental research and applications for special needs audiences should be aware that learned aesthetic values, especially those strongly reinforced in design culture, can function as unconscious processes in our work (preferences for small type and subtle variations in hue and value are just two examples). The act of entering into a research environment requires continual examination of such values and assumptions. Orientations such as these act as sources of bias in investigations in that they determine the properties of the prototypes that are created and tested. For this reason, the entire question of how prototypes are conceived and generated for communication and behavioral research needs sustained methodological development.

The potential for designers and design academics to contribute to experimental research in the social sciences is tremendous, particularly in fields responding to public health communication initiatives. Because of the increased presence of visual media and technology in contemporary culture, and resulting uncertainty as to the influence these media have on human behavior, funding for research in this area is growing as well. Preventing Drug Interactions in Older Adults provides an example of one collaboration that may lead to greater participation by designers in academic work of this kind.

## Notes

1 Some material from this article will also appear in Strickler, Z. and Neafsey, P. (2002) "Visual design of interactive software for older adults: preventing drug interactions in older adults." *Visible Language*, 36: 1.

2 The research team for Preventing Drug Interactions for Older Adults were: Patricia Neafsey, Ph.D. (pharmacology), headed the project as principal investigator. Design collaborator, Zoe Strickler, M.Des. (visual communication design), directed the visual communication research, and design and production of the animations. Collaborators Robin H. Froman, Ph.D. (educational psychology), and Steven V. Owen, Ph.D. (educational psychology), provided guidance for development of the measurement instruments and will contribute statistical

analysis for the clinical trial. Doctoral nursing student Juliette Shelman, assisted by honors nursing student Antoinette Padula, lead the qualitative language pre-study for design of measurement instruments and the program. She is also directing the field nurses who are implementing the clinical trial. Design students Michael Skiles and Mai Phung contributed design of prototypes for the formative research phase. Design students Amy Ellingham, Sam Kim, and Meena Stout provided production assistance for segments of the PEP.

3 Animation components were designed using Adobe Illustrator and Adobe Photoshop software on Macintosh platform equipment and imported into the Adobe AfterEffects software program for animation. Finished animations were transcribed to the IBM platform. The interactive program was written in the Macromedia Authorware program on IBM platform equipment.

## References


Adams, W. L. (1995) "Potential for adverse drug-alcohol interactions among retirement community residents." *Journal of the American Geriatric Society*, 43, 1021–1–25.

Anderson, L. A., Satterfield, D., German, R. and Anderson, R. M. (1996) "Using quantitative and qualitative methods to pretest the publication 'Take Charge of Your Diabetes: A Guide for Care.' " *Diabetes Educator*, 22(6), 598–604.

Bandura A. (1986) *Social Foundations of Thought and Action: A Social Cognitive Theory* (Englewood Cliffs, NJ: Prentice Hall).

Bandura A. (1997a) *Self-efficacy: The Exercise of Control* (New York: WH Freeman).

Bandura, A. (1997b) "Self-efficacy: Toward a unifying theory of behavioral change." *Psychological Reviews*, 84, 191–215.

Bandura, A., Barbaranelli, C., Caprara, G. U. and Pastorelli, I. (1996) "Multifaceted impact of self-efficacy beliefs on academic functioning. *Child Development,* 67(3), 1206–22.

Bloom, J. A., Frank, J. W., Shafir, M. S., and Martiquet, P. (1993) "Potentially undesirable prescribing and drug use among the elderly." *Canadian Family Physician,* 39, 2337–45.

Davis, T., Crouch, M., Wills, G., Miller, S., and Abdehou, D. (1990) "The gap between patient reading comprehension and the readability of patient education materials." *Journal of Family Practice*, 31, 533–8.

Davis, T., Long, W., Jackson, R., Mayeaux, E., George, R., Murphy, P. and Crouch, M. (1993) "Rapid estimate of adult literacy in medicine: a shortened screening instrument." *Family Medicine*, 25, 391–5.

Fisk, A. D. and Rogers, W. A. (eds) (1997) *Handbook of Human Factors and the Older Adult* (New York: Academic Press).

Forster, L. E., Pollow, R. and Stoller, E. P. (1993) "Alcohol use and potential risk for alcohol-related adverse drug reactions among community-based elderly." *Journal of Community Health,* 18(4), 225–39.

Fowler, F. J. (1984) *Survey Research Methods* (Beverly Hills: Sage Publications).

Froman, R. D. (1996) "An evaluation model for health sciences education projects." Invited address at the *Annual Sepa Meeting*, sponsored by the NIH, Washington, DC.

Froman, R. D., Hence, C. and Neafsey, P. J. (1993) "A comparative assessment of interactive video disc instruction." *Computers in Nursing*, 11:5, 236–42.

Gardner-Bonneau, D. and Gosbee, J. (1997) "Health care and rehabilitation." In Fisk, A. D. and Rogers, W. A. (eds) *Handbook of Human Factors and the Older Adult* (New York: Academic Press), 231–53.

Gazmararian, J. A., Baker, D. W., Williams, M. V., Parker, R. M., Scott, T. L., Green, D.C., Fehrenbach, S.N., Ren, J., and Koplan, J. P. (1999) "Health literacy among Medicare enrollees in a managed care organization." *Journal of the American Medical Association,* 281, 545–51.

Hanlon, J. T., Fillenbaum, G. G., Burchett, B., Wall, W. E., Service, C., Blazer, D. G. and George, L. K. (1992) "Drug-use patterns among black and nonblack community dwelling elderly." *The Annals of Pharmacotherapy,* 26, 679–85.

Jackson, R., Davis, T., Murphy, P. Bairnsfather, L. and George, R. (1994) "Reading deficiencies in older patients." *American Journal of Medical Science*, 308, 79–82.

Katz, S. (1970) "Progress in development of the index of ADL." *Gerontologist*, 10, 20–30.

Kline. D. W. and Scialfa, C. T. (1997) "Sensory and perceptual functioning: basic research and human factors implications." In Fisk, A. D. and Rogers, W. A. (eds) *Handbook of Human Factors and the Older Adult* (New York: Academic Press), 27–54.

Kirsh, I., Jungeblut, A., Jenkins, L. and Kolstad, A. (1993) *Adult Literacy in America: A First Look at the Results of the National Adult Literacy Survey* (Washington, DC: National Center for Education Statistics, US Department of Education).

Laubach, R. and Koschnick, K. (1997) *Using Readability Formulas for Easy Adult Materials* (Syracuse, NY: New Readers Press).

Lazarou, J., Pomeranz, B. H. and Corey, P. N. (1998) "Incidence of adverse drug reactions in hospitalized patients." *Journal of the American Medical Association,* 279, 1200–5.

Lisansky Gomberg, E. S. (1995) "Older women and alcohol: use and abuse." *Recent Developments in Alcoholism*, 12, 61–79.

Manasse, H. R. (1995) "Toward defining and applying a higher standard of quality for medication use in the United States." *American Journal of Health-Systems Pharmacy*, 52, 374–9.

Moore, A., Hays, R., Greendale, G., Damesyn, M. and Reuben, D. (1999) "Drinking habits among older persons: findings from the NHANES I epidemiologic follow-up study (1982–84)." *Journal of the American Geriatric Society*, 47, 344–8.

Morgan, D. L. (1988) *Focus Groups as Qualitative Research* (Beverly Hills: Sage Publications).

Morrell, R. W., and Echt, K. V. (1997) "Designing written instructions for older adults: learning to use computers." In Fisk, A. D. and Rogers, W. A. (eds) *Handbook of Human Factors and the Older Adult* (New York: Academic Press), 335–61.

Morris, J. M. (1994) "User interface design for older adults." *Interacting with Computers,* 6(4), 373–93.

Murdock, J. and Neafsey, P. J. (1995) "Self-efficacy measurements: an approach for predicting practice outcomes in continuing education." *Journal of Continuing Education in Nursing,* 26(4), 158–65.

Nagi, S. Z. (1976) "An epidemiology of disability among adults in the United States." *The Milbank Quarterly,* 54, 439–68.

National Center for Health Statistics, US Department of Health and Human Services (1993) *National Health Interview Service*, Series 10, No. 190.

Neafsey, P. J. (1997) "Computer-assisted instruction for home study: A new venture for continuing education programs in nursing." *Journal of Continuing Education in Nursing*, 28, 3–12.

Neafsey, P. J. (1998) "Immediate and enduring changes in knowledge and self-efficacy following home-study by computer of the pharmacology of alcohol." *Journal of Continuing Education in Nursing*, 29, 173–81.

Neafsey, P. J., Strickler, Z., Shellman, J. and Padula, A. (2001). "Use of touch-screen equipped notebook computers to deliver health information about self-medication to older adults." *Journal of Gerontological Nursing*, 27(1), 19–27.

O'Leary, A. (1985) "Self-efficacy and health." *Behavior Research and Therapy*, 23, 437–51.

Pfeiffer, E. (1974) "A short portable mental status questionnaire for the assessment of organic brain deficit in elderly patients." *Journal of the American Geriatric Society*, 23, 433–41.

Plimpton, S., and Root, J. (1994) "Materials and strategies that work in low literacy health communication." *Public Health Reports*, 109, 86–92.

Poggenpohl, S. (1998) "Growing a knowledge base." In proceedings from the conference Doctoral Education in Design (Columbus: Ohio State University), 117–29.

Pollow, R. L., Stoller, E. P., Forster, L. E. and Duniho, T. S. (1994) "Drug combinations and potential for risk of adverse drug reaction among community-dwelling elderly." *Nursing Research*, 43(1), 44–9.

Rosow, I. and Breslau, N. (1996) A Guttman health scale for the aged. *Journal of Gerontology*, 21, 556–9.

Salzman, C. (1995) "Medication compliance in the elderly." *Journal of Clinical Psychology*, 56 (supplement 1), 18–22.

Strecher, V. J., DeVellis, B. M., Becker, M. H. and Rosenstock, I. M. (1986) "The role of self-efficacy in achieving health behavior change." *Health Educator Quarterly*, 13, 73–91.

Strickler, Z. (1998) "Interdisciplinary collaboration as a means for developing doctoral-level research in visual communication design. In *Proceedings from the conference Doctoral Education in Design* (Columbus: Ohio State University), 311–29.

Strickler, Z. and Neafsey, P. J. (2002) "Visual design of interactive software for older adults: preventing drug interactions in older adults." *Visible Language*, 36: 1.

The Taskforce for Compliance (1994) *Noncompliance with Medications: An Economic Tragedy with Important Implications for Health Care Reform* (available from The Taskforce for Compliance. 2300 N. Charles Street, Suite 200, Baltimore MD 21218).

United States General Accounting Office (1995) *Prescription Drugs and the Elderly* (GA/HEHS-95-152; Gaithersburg, Md.: U.S. General Accounting Office).

Wallsten, S. M., Sullivan, R. J., Hanlon, J. T., Blazer, D. G., Tyrey, M. J. and Westlund, R. (1995) "Medication taking behaviors in the high- and low-functioning elderly: MacArthur field studies of successful aging." *Annals of Pharmacotherapy*, 29, 359–64.

West S. (1990) *Working with Style: Traditional and Modern Approaches to Layout and Typography* (New York: Watson Guptill).

Williams, M. V., Parker, R. M., Baker, D. W., Parikh, N. S., Pitkin, K., Coates, W. D. and Nurss, J. R. (1995) "Inadequate functional health literacy among patients at two public hospitals." *Journal of the American Medical Association*, 274, 1677–82.

Winkler, D. (1997) "Design practice and education: moving beyond the Bauhaus model." In Frascara, J., *User-centred Graphic Design: Mass Communications and Social Change* (London: Taylor and Francis), 129–35.

# 11 Fieldnotes from home

## Anthropology and design on exhibition[1]

*Rae Bridgman*

### *House/Home*

This article focuses on an exhibition entitled *House/Home*—an exhibition about Strachan House. The exhibition was mounted in the spring of 1999 in the Photo Passage at Harbourfront Centre in Toronto, and was part of the third annual Toronto's Festival of Photography, known as Contact '99, featuring more than 130 exhibitions and educational programs. Harbourfront Centre itself is a very high profile cultural, educational and recreational center in Toronto and presents hundreds of events and activities to the public annually. Latest attendance figures cited over 4 million visitors yearly. At the time of writing this chapter, arrangements were being made to have the exhibition tour several other galleries at design institutions, and the exhibition was to be donated to the collection of the City of Toronto Archives (Figure 11.1).

*Figure 11.1 House/Home* exhibition presentation.

The article discusses some of my reflections about *House/Home*, an exhibition of architectural drawings, photographs and fieldnotes. The exhibition represents acts of collaboration between architects, photographers, an anthropologist, and staff and residents of Strachan House. Addressed in this chapter are a set of issues that I feel are fundamental to engaged anthropological and design practices, that is our social responsibility for accommodating diverse needs and designing with disenfranchised groups. The article explores three major themes arising from my experiences of participating in this exhibition. The first looks at expanding the horizons of what engaged urban anthropology can be. The second accosts experimental ways of representing what we do as urban anthropologists and design professionals. The third considers how researchers and designers may fruitfully collaborate to share the results of their different perspectives and knowledge bases with the general public.

Strachan House offers a unique self-government form of housing that has attracted international attention since it opened in December 1996. Developed by the Homes First Society, and following from the first generation model of StreetCity, which opened in 1988, Strachan House was designed by Levitt Goodman Architects and was created to respond to the specific needs of chronically homeless women and men. In an abandoned turn-of-the-century timber and brick warehouse owned by the city, the architects designed a series of "streets" connecting "houses" to lodge 70 residents. The houses have 5 to 7 private bedrooms each and shared kitchens, bathrooms, living rooms and front porches. All the streets lead to a three-storey central space, organized around an existing and dramatic smokestack. The area is known as the Town Hall, and residents and staff meet here bi-monthly to air grievances, create policies and celebrate events. Levitt Goodman Architects recently received a 1999 Governor General's Award for Excellence in Architecture for their work on Strachan House.

Photographers Debra Friedman and Robert Burley began documenting the construction of Strachan House in the spring of 1995. Their goal was to create a photographic record of the project from the initial phases of construction through to its inhabited state (residents moved into Strachan House December 1996). Burley documented the architectural components of the project, while Friedman concentrated on making portraits (Figure 11.2). Their combined photographs not only document the structural evolution of Strachan House but also explore the ways in which the residents have created homes for themselves. The diptychs created by Friedman and Burley are meant to blend environmental portraits with a record of the developing environment itself—a technique commonly employed by editorial publications that cover "home decor." This strategy presents the home and inhabitant as inseparable (Figure 11.3).

The photographs in and by themselves provide a wealth of data to be analyzed, not a task I will undertake here. Suffice it to say that the angles and perspectives shift with each set of photographs. The two different photographs, although apparently united within one frame (an intimate juxtaposition) present different kinds of information, and evoke different kinds of relationships. Running beneath the diptychs is a series of pages from my fieldnotes (Figures 11.4 and 11.5).

*Figure 11.2* *House/Home*, an image by Debra Friedman and Robert Burley.

*Figure 11.3* View of the exhibition.

I have been documenting the life histories of StreetCity, Strachan House and Savard's since 1995 (Anderson 1997; Bridgman, 1998a, 1999, 2000).

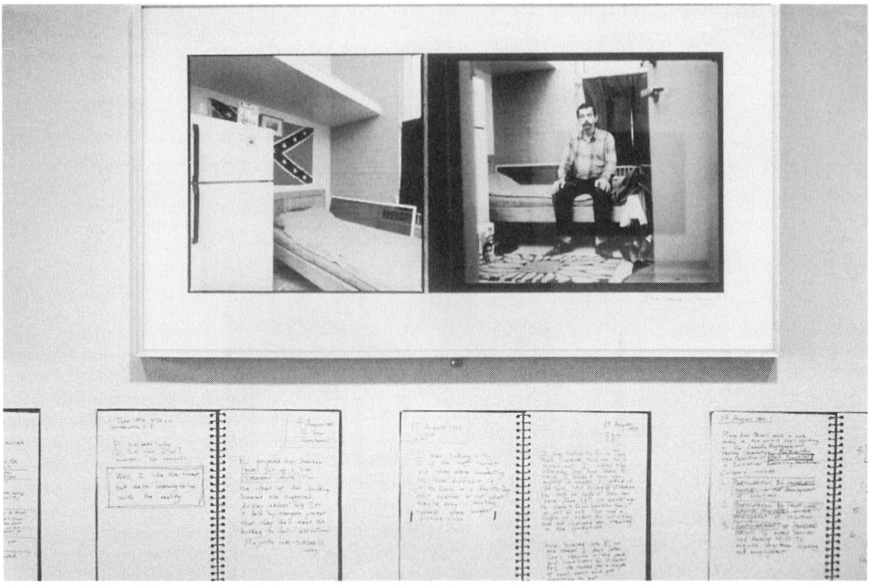

*Figure 11.4* Photos by Debra Friedman and Robert Burley and fieldnotes by the author/1.

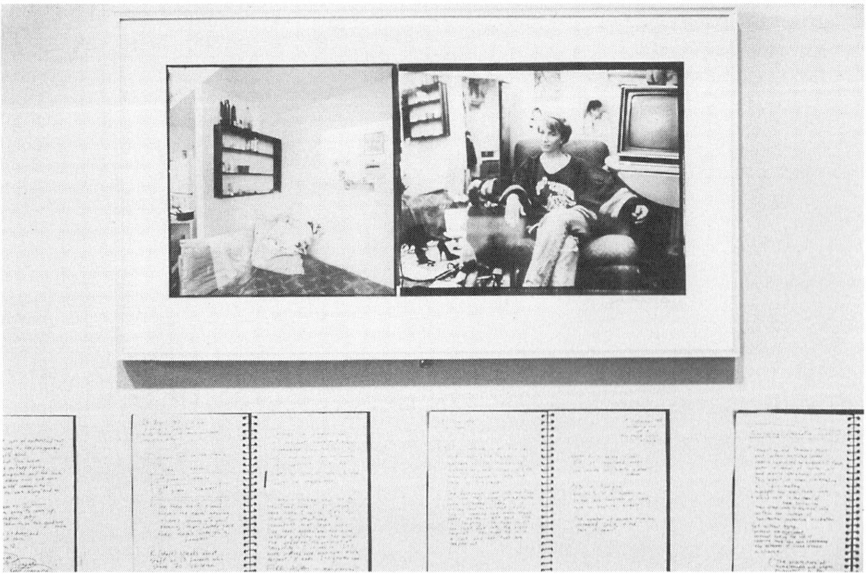

*Figure 11.5* Photos by Debra Friedman and Robert Burley and fieldnotes by the author/2.

My focus has been on the everyday processes involved in designing, developing, building, and living and working within these three innovative housing projects for those who have been chronically homeless. In *Journeys Home: Fragments from Fieldnotes, Strachan House, 1997–1998*, I drew together excerpts from my fieldnotes—archives, interviews and conversations with residents and staff, and notes taken during Town Council meetings at Strachan House. Many of the excerpts speak about different meanings associated with the word home. They also speak about balancing the needs of the individual with the needs of the community in this self-governance model of housing, attempting to provide a larger context for the place of this project within best practice models to alleviate homelessness. The notes are chronologically arranged. The effect is one of deliberate fragmentation to convey some of the processes involved in undertaking long-term ethnographic research.

## Experimental ways of representing what we do as engaged practitioners

The fieldnotes in the exhibition are not fieldnotes in the truest sense, for they are at one further refined remove from my original fieldnotes, those records of the mundane, the quotidian, the inchoate, the barely understood at the time it was written. The notes had to be written so that people could read them (my handwriting is notoriously difficult to read). They are selected, fragmentary, and are highly ordered and organized. In this sense they are not "real" "authentic" fieldnotes—I have been careful to call them fragments from fieldnotes (Figures 11.6, 11.7, and 11.8).

As a writer I have employed what have been identified as five basic strategies: pulling the reader into the story being told; recreating the immediacy of experience within the writing; including elements of surprise; reconstructing experiences through "written images"; and creating a sense of closure on the story, presenting a finished piece, a product if you will, even while recognizing the work as part of an ongoing process (Mitchell and Charmaz 1996: 144–5).

The opportunity to participate in this exhibition has drawn together my two careers, one as anthropologist, one as visual artist. I was inspired by "The Vellour/velvet/vellum Notebook" by bp Nichol (1998) reprinted in the journal *West Coast Line*—a notebook of jottings, musings, fragments of word play—it reminded me of my own fieldnotes, and pushed me to consider further how those moments I have witnessed and those excerpts from interviews could be shared through more than prose. Using residents' and staff's words—and listening carefully to their cadences, emphases, where the breath is drawn, and other rhythms—leads to poetry, leads to "touch[ing] us where we live, in our bodies, and invites us to experience reflexivity and the transformational process of self-creation" (Richardson 1996: 8).

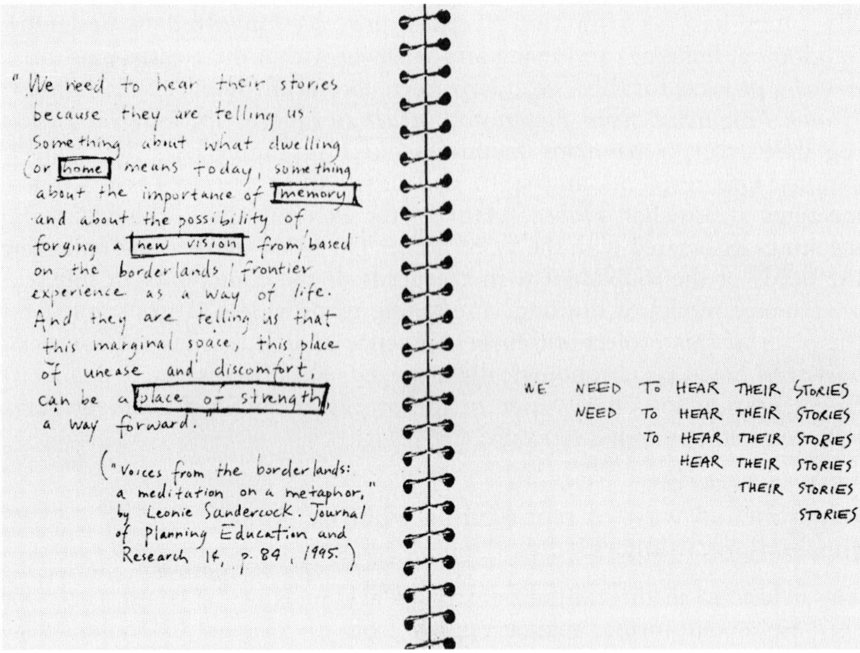


*Figure 11.6* Fieldnotes book by the author, pages 1 and 2.

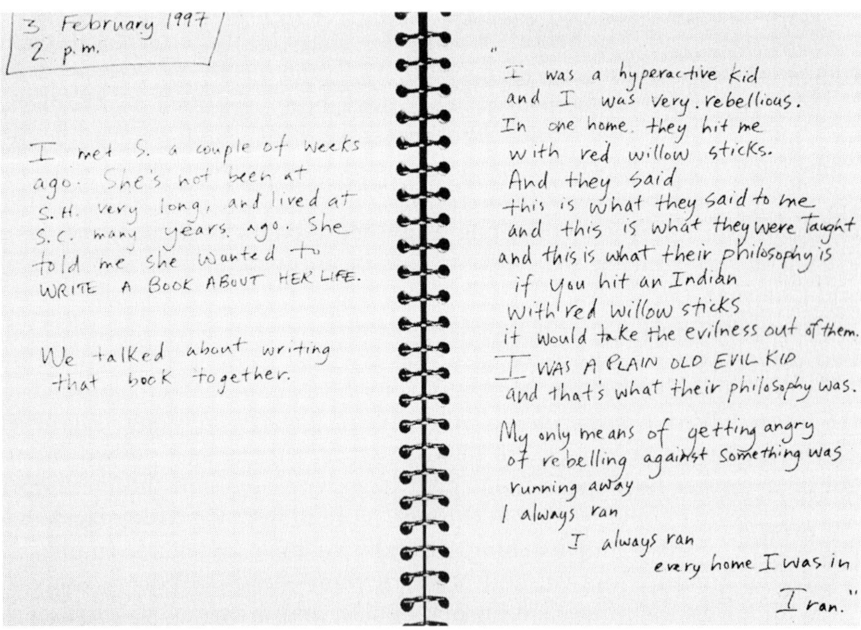


*Figure 11.7* Fieldnotes book by the author, pages 9 and 10.

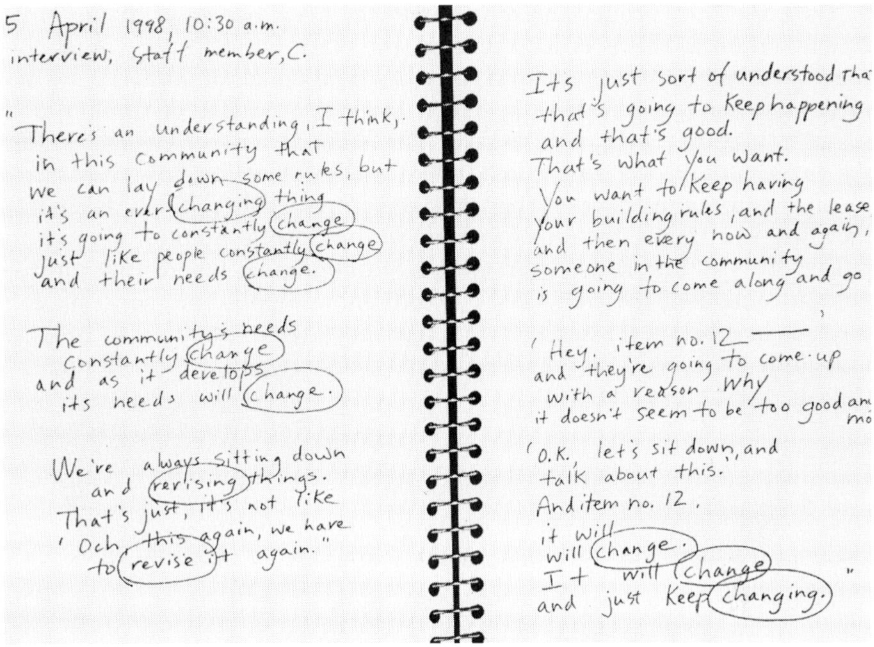

5 April 1998, 10:30 a.m.
interview, Staff member, C.

"There's an understanding, I think,
in this community that
we can lay down some rules, but
it's an ever changing thing
it's going to constantly change
just like people constantly change
and their needs change.

The community's needs
constantly change
and as it develops
its needs will change.

We're always sitting down
and revising things.
That's just, it's not like
'Och, this again, we have
to revise it again."

It's just sort of understood tha
that's going to keep happening
and that's good.
That's what you want.
You want to keep having
your building rules, and the lease
and then every now and again,
someone in the community
is going to come along and go
'Hey, item no. 12 ——
and they're going to come up
with a reason why
it doesn't seem to be too good an
mo
'O.K., let's sit down, and
talk about this.
And item no. 12,
it will change.
It will change
and just keep changing."

*Figure 11.8* Fieldnotes book by the author, pages 33 and 34.

After reading through bp Nichol's work, I realized that I did not want to present an authoritative piece of linear writing about the project. Rather, I wanted to impart a sense of what it is like to undertake long-term fieldwork, what a fragmentary and painstaking process it is to witness and to "hang out" (Rosenthal 1991).

Audience response was very good. One of the photographers spoke about how she appreciated the fieldnotes for giving her a greater depth of knowledge of the history of the project, for she had really only visited Strachan House relatively briefly. Others have appreciated the separate yet parallel echoes as the text runs under the photos, when the text does not refer directly to the photos but does connect in subtle ways.

Still others seem to have appreciated the ways in which a sense of process is shared. Most exhibitions do not share a sense of process, of how work came to fruition. This is true of most architectural design practice in which glossy pictures (most often without people) publish buildings as finished products. We seldom are able to share in the knowledge of how the project came to fruition, and how it has become actively inhabited. A key part of engaged practice should seek out opportunities for sharing these processes. The fieldnotes and photographs attempt to convey a number of difference processes at work—research processes, design processes, habitation processes, community development processes.

The photos are printed full-frame, with the black border of the negative clearly outlined. This is echoed through the black outline of the notebook pages. The full frame implies that there has been no act of cropping, no act of deletion, no act of selection, but in fact the processes of framing and selection become all the more intense. The notes require the engagement of the viewer to take the time to read them and to reflect on how they relate to the photographs above. That engagement can be relatively brief, for these are just fragments. Other viewers devote fifteen minutes or more to reflect. This is interactive ethnography/photography, where the anthropologist's mission, the architects' mission, and the photographers' mission are to help others feel a sense of "ethnographic 'truth' and thus to become more fully immersed—morally, aesthetically, emotionally, and intellectually" (Bochner and Ellies 1996: 4).

## Expanding the horizons of engaged urban anthropology and design practices

How do we share the work that we do? How can we fruitfully work with urban others on complex issues? Applied anthropology is most often understood, I think, as being an anthropology concerned with problem-solving, with recommending and acting on interventions that are meant to address perceived problems. The same could be said of design practices in other contexts. Engaged researchers and designers, I would suggest, are both concerned with working towards a better world, a utopian impulse if there ever was one.

Employment of the word "engaged" is inspired by an article by Jeff Halper and Anita Nudelman (1993) in which they distinguish between applied, practicing and engaged anthropology. Halper calls "engaged" anthropology "that borderland between practicing and applied anthropology (which have an identifiable and accepted place within the profession) and those activities which are informed by anthropological views, concepts, and concerns but go beyond disciplinary boundaries" (Halper and Nudelman 1993: 4). Engaged anthropologists explore roles played out in wider public activity.

Engagement for the researcher need not always happen through the usual means associated with applied anthropology of policy analysis and working towards change at this level. Engagement for the designer need not always happen just through the object designed. Engaged, for me, implies a larger mission of engaging multiple senses, helping others to make a shift in their perceptions of homelessness as a problem, and by corollary the homeless as a problem, who are perceived as unable to participate in working on their own behalf towards solutions to homelessness. Bringing work to public scrutiny in such a way as this exhibition does reaches a very broad audience—who may not necessarily be expecting to be confronted with such work, on their way to the local cafe to get lunch!

Engaged practice requires working actively to bring forth the voices of those who have been marginalized by society-at-large.

The last set of fieldnotes depicts a blank page (Figure 11.9)—reflecting the nature of fieldwork as a never-ending process, and the dialogue between research and design as involving an open-ended process, an unwritten page. . . .

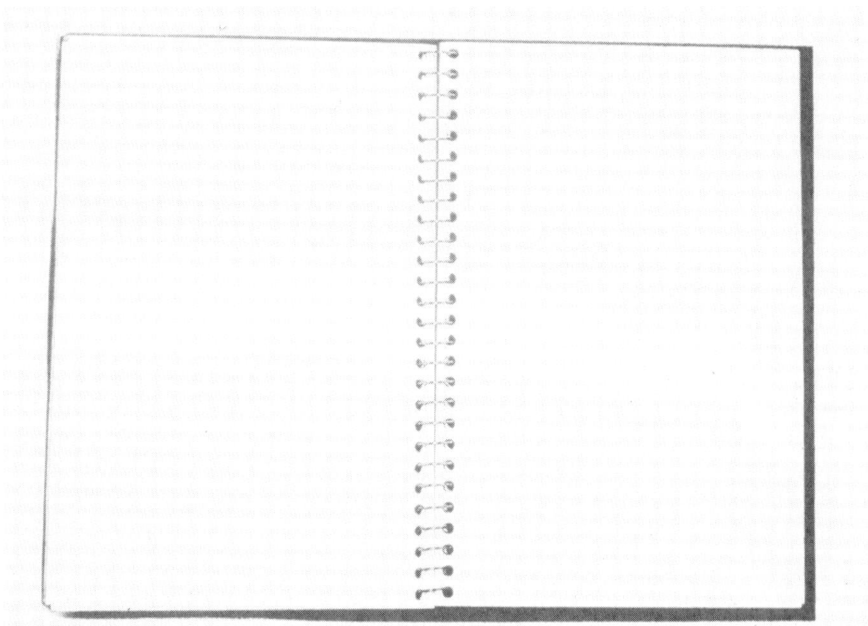

*Figure 11.9* Fieldnotes book by the author.

## Note

1 This article represents a revised version of a paper originally delivered at the annual meeting of the Canadian Anthropology Society/Socièté canadienne d'anthropologie, Universite Laval, Quebec, 15 May 1999. Gratefully acknowledged is the Social Sciences and Humanities Research Council of Canada for their support of this work through a Strategic Grant (Women and Change, 1995–1998). The photographs are by Robert Burley and Debra Friedman/Design Archive, reproduced by kind permission of the authors.

## References

Anderson, R. (now known as Rae Bridgman) (1997) "The street as metaphor in housing for the homeless." *Journal of Social Distress and the Homeless*, 6(1), 1–12.

Bochner, A. and Ellies, E. (1996) "Taking ethnography into the twenty-first century." *Journal of Contemporary Ethnography*, 25(1), 3–5.

Bridgman, R. (1998a), "The architecture of homelessness and utopian pragmatics." *Utopian Studies* 9(1), 60–7.

Bridgman, R. (1998b), "A 'city' within the city: a Canadian housing model for the homeless." *Open House International*, 23(1), 12–21.

Bridgman, R. (1999) "The street gives and the street takes: the meaning of the 'street' for those who are homeless." In Williams, A. (ed.) *Therapeutic Landscapes: The Dynamic Between Place and Wellness* (Lanham, Mass.: University Press of America), 153–66.

Bridgman, R. (2000) "My journey home: homeless and Aboriginal and a woman." In Oakes, J., Riewe, R., Koolage, S., Simpson, L. and Shuster, N. (eds), *Aboriginal Health, Identity and Resources* (Winnipeg: Departments of Native Studies and Zoology; and Faculty of Graduate Studies, University of Manitoba), 93–104.

Halper, J. and Nudelman, A. (1993) "Applied, practicing and engaged anthropology in Israel." *Practicing Anthropology*, 15(2), 3–4.

Mitchell, R. and Charmaz, K. (1996) "Telling tales, writing stories: postmodernist visions and realist images in ethnographic writing." *Journal of Contemporary Ethnography* 25(1), 144–66.

Nichol, bp, (1998) "The Vellour/velvet/vellum Notebook." Reprinted in *West Coast Line*, 25, Spring/Summer.

Richardson, L. (1996) "Educational birds." *Journal of Contemporary Ethnography*, 25(1), 6–21.

Rosenthal, R. (1991) "Straighter from the source: alternative methods of researching homelessness." *Urban Anthropology*, 20(2), 109–26.

# 12 Research and design collaboration

## A case study

*Louise St Pierre*

## Introduction

This chapter documents a case study where I employed a researcher to conduct user observation of a product prototype that I am designing. Aside from the hope for new research findings, it was my intent to explore experientially the differences between a designer's and a social scientist's methodology.

Designers often practice research and observation, but do not have a strongly established methodology. It has been common to go out there (into the real world) and find things out for yourself, feel the other point of view, live it, experience it, see what it feels like. My term for this intuitively based process is "empathic immersion." The anticipated outcome of this immersion is often a sympathetic attunement of the design in response to a new understanding. More ambitious design research includes interviewing users (listening to their words) testing their use of prototypes (observing their actions), and responding to the knowledge gained by this with appropriate modifications to the design.

Research work in the industrial design profession, however, is becoming increasingly sophisticated, with firms often hiring sociologists, psychologists, and anthropologists to provide base studies as a framework for product development. The burgeoning discussions about cross-disciplinary activity focus on how different disciplines can work together most effectively.

As the researcher, Courtney O'Catherine, and I worked through the case study, preconceptions about each other's roles, disciplines and abilities came to light. In the end, we found ways to work together which transcended disciplinary boundaries and allowed the project to dictate what was most necessary.

## Chronology of the study

The opportunity for this comparison was my ongoing study of a group of preschool children aged 4–6 as they interacted with a series of furniture components that I had been developing. I isolated a specific period of time

(Phase One: Spring/Summer 1998) when I had very actively been doing observation. A similar opportunity was created (Phase Two: Spring/Summer 1999) for the researcher to take primary responsibility. The study facility, a parent run co-operative preschool, remained constant through both phases.

### *Phase One: Spring/Summer 1998*

During this phase: (a) I was well connected with the study facility, as my own daughter was in their daily care program; (b) as a working parent in a co-operative, I was comfortable working in the classroom, and knew the routines and expectations; (c) I enjoyed a friendly working relationship with the teachers; (d) I had a warm connection with the children; (e) staff were involved in the development of the project under study.

Over the period of Spring and Summer 1998, a series of furniture components were installed in the test facility (table, bench and screen), and I observed how the children used them. My primary tools were camera and video. I asked direct questions of the teachers and the children, and modified the initial design based on their responses. An example of this response to direct feedback is shown in Figures 12.1 and 12.2. In Figure 12.1, the child is asking "Why doesn't this stack up right?" After asking the teachers if they felt comfortable with having the components stack in order to be used as building blocks by the children, I revised them so that they could do so (Figure 12.2). The resulting kit of parts allows children to shape their own spaces, depending on the comfort level of the staff.

During this phase, I experienced observation as an intuitively based, immersive activity. I did not reflect on myself as an influence in the observation process, though I often wondered if people were giving me positively oriented answers because they knew me. Most of the time, I did not presume that I was working with any great amount of objectivity.

### *Phase Two: Spring/Summer 1999*

During this phase: (a) My child no longer attended this facility; (b) the staff had undergone a complete turnover; (c) many of the children were unfamiliar to me; (d) I was not present on the site very often; e) Ms. O'Catherine, who had strong experience in education, conducted all the research; (f) staff felt little or no commitment or ownership of the project, however they were happy to accommodate Ms. O'Catherine on the site.

Phase Two of this project is significantly different from Phase One in that the new components being tested, display ladder, tray and box, are not as familiar to use as the table, bench, and screen which were tested in Phase One. These new items were an attempt to test the ability of children to display their classroom activities in a more comprehensive manner than that which is normally done in North American daycares. The inspiration

*Figure 12.1* "Why doesn't this stack right up?"

for this experiment is the preschools in Reggio Emilia, Italy, where staff members are responsible for documentation and presentation of children's activities. This is done by including audio recordings, photographs of children working, and samples of the children's work. This type of comprehensive display has been shown to increase children's ownership in learning as well as parental involvement (Gandini 1993: 146). In North America, where we have less financial support for preschools, and fewer staff, comprehensive display is rarely possible. The display furniture was developed as an experiment to test the hypothesis that, with the right design, children might be able to participate in creating displays of their own work, thus accomplishing similar goals as at the Reggio Emilia Schools, without the same kinds of demands on staff.

*Figure 12.2* Stackable components in use.

Knowing that I was assessing our working relationship as well as hoping for some insight into the use of the new furniture components, I gave Ms. O'Catherine little direct instruction aside from a thorough introduction to my design goals. While I was deliberate in not over-directing her, I did struggle at one point to try and define the problem I wanted her to look at: "I am really struggling to find the right question to ask. What I really want to know is perhaps too general: Does this piece of furniture make a difference? Does it change the way the children work? Can they use it easily? Does it make intuitive sense for them? Does it change the way adults see and genuinely understand their work?" (email correspondence, 27 February 1999).

This struggle is informative of itself; it feels limiting to the design to try to isolate a single question that can then be accurately researched and measured by social science methodology. I had many questions, and assumed many more could arise during the observation itself. I also had no experience in setting up a structured observation, and worried that a tightly defined research process might limit the open endedness that I was comfortable with.

Ms. O'Catherine began with baseline observations and interviews before the new furniture pieces were installed, and then conducted observational research (photographic documentation) of the pieces after installation. Despite a number of different locations for the display ladder over the ensuing weeks, it did not draw much interest or use from the children and teachers. Ms. O'Catherine grew increasingly frustrated, as there was little to document. I began to feel an uncomfortable lack of control over both the research methodology and how the prototype was being used. After revisiting the site, I saw that the display ladder was being used for storage of unwanted children's art (and in fact looked terrible). I did what I had always done: I acted. This time, I acted by clearing all the shelves in the hopes that there could be a new beginning and the teachers and children would work together to create meaningful display.

This, unfortunately, upset Ms. O'Catherine's data. I had introduced too many variables by making these wholesale changes. Fortunately however, it represented a turning point in the project where we finally felt liberated to design the observation activity that the project needed (see Figure 12.3, mid-June 1999).

At our subsequent meeting, I learned, much to my surprise, that Ms. O'Catherine did not claim to be any more objective than I had in Phase One. She in fact felt that she was barely able to hold herself back from telling the staff and children how to use this display ladder, and desperately wanted it to work and be fully accepted. Seeing it neglected was frustrating for her. Further, she felt strongly that there was a flaw somewhere in the initial premise that display was innate to children. "Using your other furniture is innate. Building and constructing is innate. But display is not innate," she said. Ms. O'Catherine secured my agreement to demonstrate the potential of this piece of furniture by running a series of structured activities, which she termed interventions. She prepared a specific program, had it approved by the staff, and over the next few weeks, took charge of the art and science activities of the children.

Under Ms. O'Catherine's direction, the display ladder was moved to an optimal location and the children participated in several activities which made full use of the display ladder and its components. During this intensive period, she was on site constantly, documenting immediate response on the part of the children, progressing through to the parents' increasing interest level, and finally, full commitment of the teachers. She noticed specific types of functionality for the display ladder, such as parents using it as a transitional tool to re-engage their children in preschool at morning drop-off. And she noticed and documented children re-visiting their work, their greater meta-cognitive process, and new actions in sharing their activities with younger siblings.

One parent noted: "We have always seen our daughter bring home tons and tons of artwork, but this was the first time we could see what she was thinking about while she was at school" (transcribed from parent interview, 27 August 1999).

During this time, Ms. O'Catherine reported to me briefly once a week. At the end of the interventions, she met with me to hand over her extensive photographs, and a final report, before leaving the state to take her new job. The compilation and assessment of the visual data was left for me.

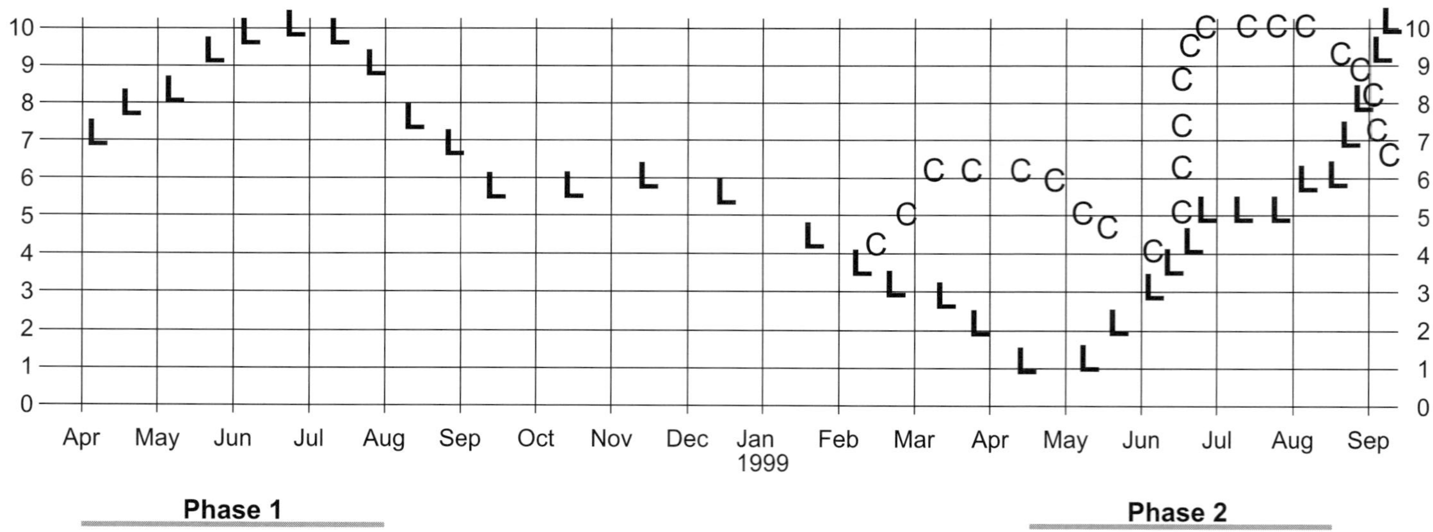


*Figure 12.3* Levels of engagement, April 1998 through September 1999.
*Note:* LLL = Louise St Pierre; CCC = Courtney O'Catherine.

## Discoveries

### *Engagement in the project and connection to the user*

While my level of engagement, or degree of occupation with the project activity (see Figure 12.3) varied throughout the duration of the study to a low of 1.0, I was able to effectively re-engage, without returning to the site, by reviewing Ms. O'Catherine's documentation materials. The photographs and oral discussions with the children and parents were compelling enough to draw me into a relationship with these people, even though I was not directly encountering them.

This relationship with the user was different from the one I had previously experienced. Despite my involvement and clear connection with the children during Phase One, I had been attempting to keep an emotional tone from entering my observation and work. Ms. O'Catherine had no such reservations, and her empathy and attunement to how the children actually felt, understood, and learned, permeates her materials, giving a resonance to her observations which is not present in my own.

### *Modifying the object or the behavior*

One of Ms. O'Catherine's more profound comments at the end of the project, was that she was surprised to learn that we were assuming that the use was "correct." She was more accustomed to entering a study with intent to modify the user's behavior, and it had not occurred to her that we might instead be intending to modify the product. This is an important distinction. The designer who wishes to design a product for easy and inuitive use will want to work with existing patterns of use as guidance for the design.

Once Ms. O'Catherine understood this, she began to offer very insightful feedback based on accommodating the needs of the children and staff. She also began to feel less concerned about our lack of a specific objective, and began to enjoy participating in the evolution of the design.

### *Respect for the vernacular*

Ms. O'Catherine taught me to accept the user's visual vernacular (see Figure 12.4). While using charming folk-image graphics may be common to preschool teachers, it is not to designers. I learned that I could not change the way visual materials looked in the classroom, and that the display should accommodate the existing vernacular.

Related to this was the information that I should not ask teachers to use new materials, such as lightweight mounting board. They need to continue to work with media they are comfortable with, such as colored craft paper.

*Figure 12.4* The visual vernacular.

### *Object vs. human focus*

Before beginning this case study, one of my strongest preconceptions was that Ms. O'Catherine would be less successful than I was at taking photographs. The base of photographic data she collected for me disproves this notion. In fact, she took some very lovely and highly descriptive photographs (see Figures 12.5 and 12.6).

She did, however, take the photographs with a different intent and sensibility than I had. I was always looking for "the one shot," that single quintessential image which would validate and explain my work. I took care to select angles so that the design was framed to look its best. There was a definite object-focus.

Sifting through Ms. O'Catherine's visual database, I realized that she was working from a human focus. In her images the children are central, rather than the products, and further, she was taking sequences of images (example shown in Figure 12.6) rather than looking for the single image. These sequences clearly indicate user activity as the primary content, rather than the object itself.

## Summary

John Zeisel says, "Possibly the most rewarding procedures to use are ones which team members jointly design to do throughout a project. Such procedures might be called 'transdisciplinary' because the criteria the team uses neither wholly reflect any one discipline, nor join different disciplines. They are new procedures developed by team members who respect each other's disciplinary norms, rewards, and sanctions, and who are willing and able to reevaluate their own norms in light of the team's common goals" (Zeisel 1984: 53).

The point at which both Ms. O'Catherine and I threw out our initial premise and worked together to deal with what the project needed at that moment, was the point at which the process was transformed. She was more comfortable embarking on an activity in which she could be her full self: an educator, a participant, and an observer. This authentic role liberated her to bring her unique perspective to the project. With the success that this approach immediately engendered, I felt more able to trust in what

*Figure 12.5* Child at work. Photo Courtney O'Catherine.

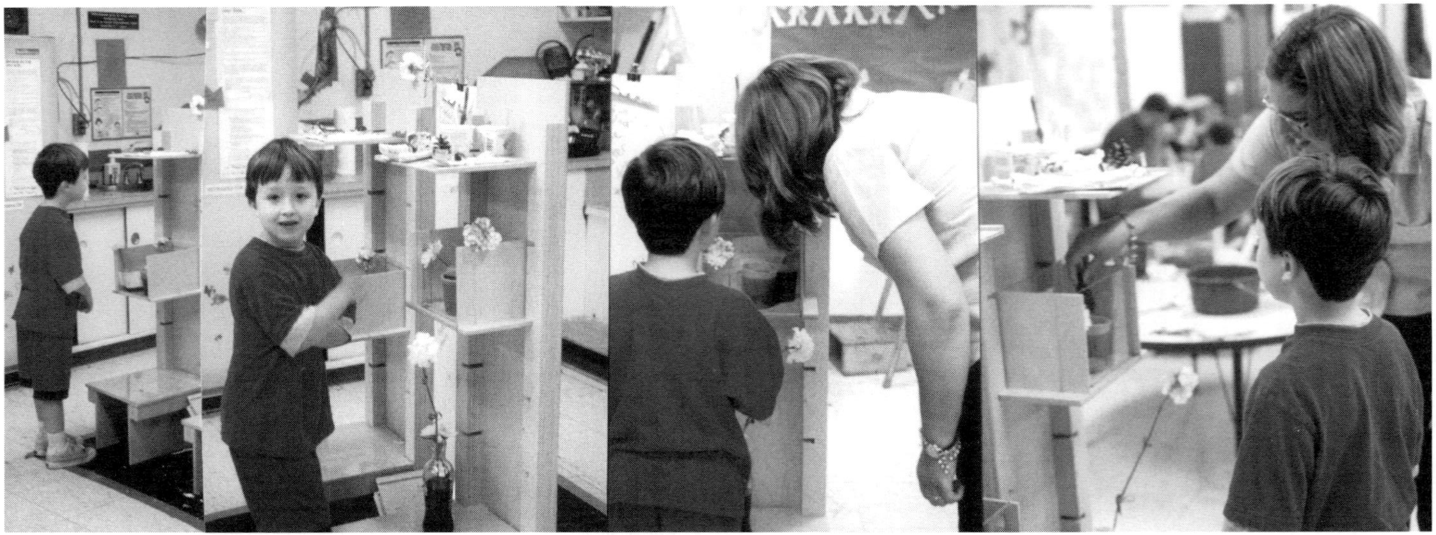

*Figure 12.6* Work sequence. Photo Courtney O'Catherine.

was happening, and allowed myself to distance from the project. It was a happy discovery that withdrawing from the project long enough to be able to re-enter through someone else's eyes offered new perspectives for the design. In the end, the design process and the product were enriched and informed by the collaboration.

## References

Gandini, L. (1993) "Educational caring spaces." In Edwards, C., Gandini, L. and Forman, G. (eds) *The Hundred Languages of Children* (Greenwich, Conn.: Ablex Publishing), 135– 50.

Zeisel, J. (1984) *Inquiry by Design* (Cambridge: Cambridge University Press).

## Acknowledgement

The Royalty Research Fund at the University of Washington provided support for the research project reported in this paper.

# 13 The technical and the social in engineering education

*Maurits Ertsen*

## Introduction

We are surrounded by the products of engineering design. Two types of properties characterize these products, and the separation between them is crucial for understanding designed artifacts (Meijer 1998). First, products have intrinsic properties: physical, chemical and biological features. Each product can be measured, the material can be defined, etc. But in design the relational properties are much more important. The products of designers are meant for something, they have goals and functions. The relational design functions are indissolubly connected to their context. The function of a bridge only exists by the grace of a context with cars, trains, roads and travelers (Meijer 1998), and a river . . . Designers create functions, which are possibilities for human action. This human action component belongs to the social sciences domain. At the same time artifacts and structures are used with properties that belong to the natural sciences domain. A design (and design education), however, is more than an addition of loose parts from both these domains. If one would bring together an economist, a traffic expert, a jurist, a physicist and a chemist, the result would not be a bridge (Meijer 1998)!

A discussion on engineering and design is essentially a discussion on problems and problem solving (Krick 1969). As a problem-solver, an engineer or designer has to judge and integrate knowledge from different perspectives and to bring them together in a synthesized design or solution. Typically his/her problem begins with the recognition of a need or want that apparently can be satisfied by some physical device, structure or process. The engineer's prime task is to translate a loose statement-of-what-is-wanted into the specifications for a satisfactory means of fulfilling that objective (Krick 1969). Many times designers are confronted with wishes that have not been translated yet into technical or designing terms (Ertsen 1999). In such a case, they have to participate in defining the problem first, before they can start working on it. This requires knowledge from different perspectives, which should also include societal aspects and socio-economic sciences. Notwithstanding the number of different opinions on the subject, something like a new consensus on the usefulness of socio-economic sciences for engineering and design appears to grow.

Introducing societal components in design asks for a specific approach to the activity of designing, which will be explained and illustrated. Designing is not a reproductive activity based upon recipes, but a process in which demands, methods and solutions have to be linked to each other in a continuous iteration and interaction with other actors. As a first step, I will discuss what, in my opinion, is a key ability of a good designer: the art to translate problems that in practice are basically unspecified into an adequate and fruitful specific design task. I will elaborate on this general idea of design capability. Examples from engineering education at Delft University of Technology or Wageningen Agricultural University are presented in boxes. Some concluding remarks about design and social sciences education in the context of the engineering curriculum are made in the last paragraph.

## Designing as the art of translation

The term "design" is applied in many engineering disciplines. Knowledge of the object to be designed would be discipline-specific, but knowledge and organization of the process appears to be generic (De Jong 1999). Whatever engineers may be creating—a nuclear power generator, dam, printing press, food-processing plant, or mechanical heart—they do so by means of the same basic design process (Krick 1969): the activity in which a problem (often in the form of a request or plan) is to be transformed from loose data into (a plan for) an artifact or other (material) structure. It involves, in some way, the following steps: problem analysis, generation of alternatives, selection and specification of one or several alternatives, testing and improving the design, and presentation of the results (Van Keulen 1998). This chapter is based on my own experience with design and design education in agricultural and civil engineering, and water infrastructures in particular. I would claim, however, that the design approaches in these fields have much in common with other fields, and I will give examples which support this claim. I invite the reader to find other common features with their own design field(s).

In practice, engineering problems do not present themselves in a prestructured way. When a problem arises, it is not directly clear how this can be formulated most adequately in engineering terms (Ertsen 1999). The notion that practical problems are basically unspecified does not imply that they could not adequately and fruitfully be translated into a specific question. Engineering practice shows that a translation in "engineering language" enables the design of many adequate solutions. At the same time, many examples show the limitations of a one-dimension approach, especially when problems become more complex. Besides technical knowledge, such problems demand input from fields such as economics, law, organizational analysis and sociology. As modern engineering problems in general involve many people and require negotiation and networking, communicative and social skills are necessary. Taking into account societal demands and conditions implies that each design should be tuned to the situation under consideration: a designer cannot come up with a standard solution. Paradoxically, introducing such a multifaceted approach asks for engineers with even higher technical qualifications. The technical knowledge of an engineer, and the way an engineer is able to command this knowledge, should be better.

## Design capability

The concept of "design capability" (Van der Ploeg 1991) is defined as the art to transform specific circumstances and problems with the help of scientific insights into new solutions. Since both the context and the scientific issues relevant for a problem are variable, one can state that this design capability can result in new technologies every time it is employed. The trend (in agricultural engineering, but probably also in others) has been rather the opposite: once a specific technology has been designed and constructed, this technology starts to speak its own "language." It starts to figure as the latest and thus the best solution for all sorts of problems. The design capability that is used is "frozen." In case a technology does not fit too well in other situations, attempts are made to change the situation in such a way that the technology becomes applicable. Development aid is a notorious example of this approach.

The capability to design, the art of designing, is disappearing from universities and polytechnics. The engineering studies become manufacturing agencies for blueprints and experts to apply them. One of the remedies appears to be to replace the blueprint approach for a systematic and generally applied concept of redesigning; existing technologies are sources of inspiration, not blueprints (Van der Ploeg 1991). In practice it will become clear that redesigning is pre-eminently an iterative process, as it is repeated until the best solution (the highest possible correspondence between conditions and design) for all actors involved is found. Such design capability and the abilities to establish a set of relevant conditions from a complex of social, economic and technical variables in an interactive process with the parties involved, are decisive qualities for designers. This asks for a renewal of design education.

## Criteria and effects

Designing is a process in which criteria, assumptions and information have to be adjusted repeatedly. It is also a cumulative process as experiences from other design contexts are used in new situations. Most design methods involve some kind of defining the desired future state of the object to be designed. Usually, design criteria or demands are derived from this definition. This is useful, as it enables designers and other interested persons to control and judge the further design process. Criteria define the (environmental and social) boundaries of the design(s). Each time a decision has to be taken, one can use the criteria as back up. This does not mean that the criteria are objective and neutral, or that "facts" could never be changed. On the contrary, criteria enable designers to make decisions during the design process, because the designer has taken the trouble to sketch beforehand the desired situation. Future decisions can be clarified by referring to this sketch. Naturally, it is not possible to define criteria without some knowledge of the future construction or system to be designed.

In practice, designers will define most of the criteria in the phase in which some preliminary studies and designs are made. Each successive step in the design process increases the finality of the design, which corresponds with a decrease in freedom in the design as more and more design conditions/criteria have been translated into set design components. The design space decreases throughout the process. In general, the design moves from rough to detail. In the beginning, the main design components are defined, implicating that decisions taken at an earlier stage are more important than those taken at a later stage. Standard design procedures usually do not include a mechanism to link the set of criteria systematically to the effects a design result can have. Certain effects will certainly be discovered and anticipated, but a systematic approach seems to be missing.

**Box I: Irrigation Design Project, Wageningen Agricultural University**

Traditionally, technical courses have constituted the most important part of the courses at the Department of Irrigation and Soil and Water Conservation (DISWC). With the upcoming criticism on irrigation development and the growing awareness of the importance of non-technical components for engineering education, social science courses were introduced as well. The design project discussed here was organized as a combination of technical courses (on structures, photo-interpretation, etc.), with some specific social sciences courses, mainly to support the analyses students had to make (Blom 1982). Groups of students analyzed an area (like Senegal) and had to design an irrigation system. A farming system analysis (FSA), in which the agricultural production system is studied, including parameters like crops, water use, labor, soil type, etc., served as link between general studies and design activities. The design course is a simulation of the process of defining criteria, making of assumptions, sketching a design, adjusting criteria, etc. Not only physical/technical criteria are developed, but also criteria of a different (socio-economical) character, which appear to be decisive too for a successful design. Irrigation is regarded as an activity in a context, and as a factor in regional developments. Much attention of the students was devoted to the analysis of the relations between design and context, acceptance by users, and criteria and choices to be made. Feedback on consequences of choices, impact of the design and related questions were lacking.

**Box II: Civil Engineering Project Education, Delft University of Technology**

The Working Group on Civil Engineering Project Education organizes about 125 projects each year, which are carried out by groups of students. In project groups students analyze a problem and design a solution. During the project all relevant aspects have to be taken into account, both social and technical, as in real practice. Depending on the year in which the project is carried out, the accent is put upon a infrastructural planning, spatial or constructive elaboration of the subject. In the second year project course a spatial and constructive design has to be made: a construction has to be designed in its direct physical environment. Parts of the construction have to be computed and dimensioned. Half of the time devoted to the project is reserved for the constructive aspects of the design, the other half for more general design aspects. One of the general items is an analysis of the effects of the construction on the environment. It appears to be difficult for students to link effects to the designed constructions. Students are able to name many potential effects, but usually these are enumerations of possibilities and not the result of a systematic study of the properties of the construction in relation to its environment. To improve the effects' analysis, teachers from the faculties of Civil Engineering and Technology, Policy and Management have started to develop a suitable methodology for an effect analysis. The effect analysis methodology uses a step-by-step approach (Ertsen and Heijer 1998):

1 determination of specific factors relevant for the situation and the design;
2 description of the desired end-situation per selected factor: design criteria;
3 analysis of the development factors over time including potential interaction;
4 determination of possible improvements and/or changes in the design.

## Designing as a learning process

Scheer (1996) introduces a model in which the design process is conceptualized as a learning process (adapted version in Figure 13.1). In the model, several stages of the learning process are made explicit. The separate learning cycles of engineers and users are linked to and confronted with each other. During the confrontation, learning experiences of both sides are shared, providing a basis for the joint knowledge that is required for quality design. In interactive design processes (future) users of the artifact or other product to be designed are involved explicitly in the decision-making process. Organizing an interactive design process is not an additional burden for projects and agencies, but recognition of actual processes, in which users have a responsibility, they will often do things with the designs that designers would not have expected. Attention should be paid to broadening the designer's frame of reference as well as the users' in order to develop effective interaction (Meijers 1990).

This lesson, which is rather manifest in irrigation engineering, can gratefully be employed in other engineering disciplines (Ertsen 1999). A recent approach, Simultaneous Engineering (SE) (Concurrent Engineering [Herder 1999; Payne *et al.* 1996]), has many similarities with interactive design. The aim of SE is to avoid as much as possible changes at a relatively late stage of the design process, as making changes in this stage is much more costly. At the same time, the danger of making mistakes at an earlier stage should be avoided. This is done by bringing the expertise that is traditionally used in later stages (like servicing, sales, component supply) to bear at the same time, early enough to resolve design and manufacturing concerns before production requirements are fixed and equipment is ordered (Payne *et al.* 1996). The concept of SE was developed in manufacturing industries like the automotive industry, aircraft and electronics. Its general applicability, however, has been increasingly recognized in other fields of engineering, like chemical engineering (Herder 1999) and civil engineering (Payne *et al.* 1996). Two factors appear to be significant for success of simultaneous engineering (Payne *et al.* 1996) and interactive irrigation design (Ertsen 1999):

1 Coordination between design and construction. If all the design work precedes all the construction work, the project must take longer than if some of the work can be done simultaneously. If the constructor has no involvement in the design phase there is no check whether the design is buildable.
2 The degree of involvement of other actors. Usually, only a small part of the expertise that is required will be available within the design team. An important early step is recognition of this outside capability by the internal experts.

An approach like simultaneous engineering or interactive design asks for a setting in which the different actors can learn from each other. People should be willing to discuss their inputs in the process, and people should be open and have a spirit of partnership. For many companies and agencies this requires a complete change of culture or style, which may take years. In the structural setting of the design process in commercially oriented (like manufacturing of consumer products) or governmentally oriented practices (like construction of infrastructural facilities) learning processes are not easily realized.

Designers have an extra role in interactive processes: besides bringing in technical expertise, they should also facilitate discussions and decision-making by the actors involved. This does not change the technical responsibility of designers. Designers remain responsible for the quality of the design, and should accept that responsibility. Interactive design approaches demand designers that are better prepared in communicative skills, but in my opinion also in technical knowledge and design experience. In contrast to traditional design approaches, in which standard solutions are selected from a limited number of available options, interactive designing brings up original plans and options, which need to be considered technically. Interactive designers need creativity and technical abilities to deal with a variety of demands, criteria and options: in sum, extended design capability.

**Box III: Role-playing, Wageningen Agricultural University**

The course in which the role-playing took place focused on the questions of how components from the social context can be translated into designs, and what adequate procedures are available to involve different actors. Role-playing enabled the students to perceive the different positions and interests of the actors involved, possibilities and constraints in interaction, and provided a safe training opportunity to deal with them. The students are provided with a setting for experiential learning. Such role-plays can be made more or less "real": using cloths and equipment, or making real physical barriers between parties who would be separated in reality. In a way many design projects that are used at technical universities are structured as a kind of role-play, in which a teacher from the faculty acts as the client. The role-play was set in a Senegalese village, which is visited by a delegation of experts to review the possibilities for the construction of a new irrigation scheme. The students playing this delegation have 45 minutes to think over how they would want to structure the visit. Then they visit the "village." In the "village" several actors are defined, such as the village chief, the president of the already existing irrigation system, a male farmer, a female farmer, etc. Each student-actor receives general information about the village structure (the division of power, for example), and specific information about his or her own role. During the visit each actor will try to play his or her role as convincingly as possible, taking into account the position they have in the village and their interests.

**Box IV: Multimedia Design Project, Delft University of Technology**

At Delft University of Technology a "Multimedia" course has been taught for some years now (Van der Mast). Student groups develop a multimedia product for a real, external client (company or other organization) and have to go through the complete trajectory from order to delivery and acceptance of the product. The main goal of the project is to learn to design in a multidisciplinary setting. An important second learning goal is to learn how to co-operate. The course focuses on the learning of skills, not on theoretical or factual knowledge. The design process of the groups is pre-structured:

Phase 1: Analysis. Based on interviews with the client the team describes the problem and formulates a program of criteria and demands. The students familiarize themselves with available software and hardware, and the available material of the client. A user analysis is made too.

Phase 2: Design. The teams define and create a solution in the form of a scenario or story line, a representation of the whole design and a detailed storyboard with sketches of all the main screens to be designed and the interactions used. Finally, some interactive prototypes showing the use of the media and a final set of criteria and demands are required.

Phase 3: Realization. The storyboard has to be realized as completely as possible. The teams have to divide the tasks and integrate the different parts at the right time. Teams and clients participate in a short acceptance procedure, after which the final product can be installed on a CD-rom.

## Discussion

Design projects are a valuable part of engineering education, as they provide students with a simulation of their future professional context. Other elements in the curriculum should be supportive by enabling students to learn what they could apply when designing. The program of Industrial Design at Delft University of Technology shows such a systematic connection between projects and lectures (Van Keulen 1998). The Aerospace Engineering and Marine Engineering programs show a comparable approach. Designing flying and floating objects asks for profound knowledge of mathematics, mechanics and materials science. These disciplines constitute a large part of the curriculum, and could be regarded as design-oriented (Van Keulen 1998). There is no reason to assume that such a general approach would be different for the social sciences.

Integration of social sciences in a curriculum, however, is not just a matter of offering the students extra courses. Students already complain that they do not know why they have to take certain courses, teachers complain that the students seem to have forgotten what they have learned in the earlier years of their study (Van Keulen 1998).

Attention for social issues should be integrated in design courses, providing a foundation for specific courses. In Box V, I give some examples of keywords that are employed in the example design courses. It is not difficult to extend this list, showing both the "natural relations" between

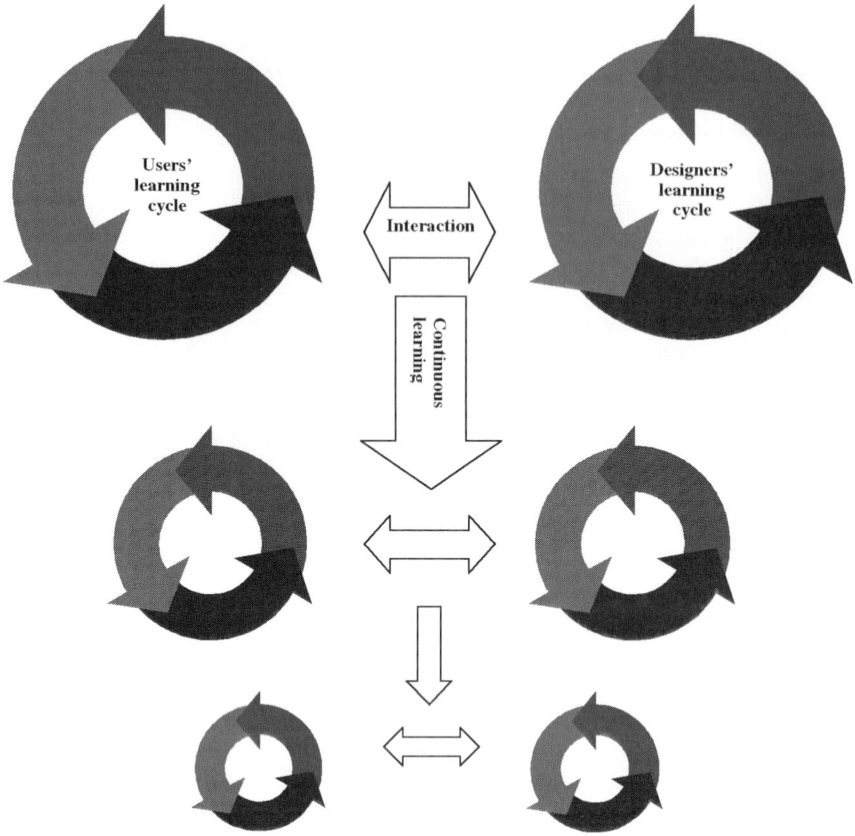


*Figure 13.1* Designing as a learning process (adapted from Ertsen 1999 and Scheer 1996).

design and society and the importance for a specific focus for the contribution of the social sciences. As design is about interfaces, a most promising issue of interaction (or interface) between design and social disciplines would be communication. People participating in design processes should be willing to discuss their inputs (values). They should understand each other (speak the same language). Every participant has a different role to play, depending on his/her position in the organization, but also on social values, power relations and daily behavior.

**Box V: Some social science keywords in interactive designing**

Behavior/communication/knowledge system/organization/rules/demands/equality/language/power relations/values/choices/groups of actors/media/roles/wishes

The conventional way of teaching the integrative design activity is to offer "learning by doing" design courses, in which students are taught a kind of design methodology. In the Civil Engineering projects (Box II) methodologies are very important, and much attention is paid to them. The Multimedia design course (Box IV) provides students with a well-described step-by-step approach, which could also be considered a design methodology. I do not deny the potential strength of design methodologies and would encourage design educators to present their courses in a well-structured way. What is important, however, is that students should be able to generalize and, if necessary, to adapt the particular methodology to other contexts as well. The methodology should not become a blueprint, but be a source of inspiration.

Learning the necessary problem-solving skills will not be sufficient for individuals to become successful problem-solvers, nor guarantee that they could tackle problems of all sorts. The process of problem-solving has some very un-skill-like characteristics (Norman 1988). The group of students from the first year of medical school studied by Norman, used exactly the same process as experts. The main difference between expert clinicians and students, however, was that experts generate better hypotheses, which is not a characteristic of a skill. The correctness of the specific hypotheses, rather than any process variable, was the strongest predictor of success. The experts are experts because they have extensive experience, and can apply their knowledge and skills to the solution of a problem. This notion corresponds well with the conception that designing is a cumulative process as experiences from other design contexts are used in new situations (also an aspect of redesigning). It does make sense for students to learn the prerequisite knowledge in the context of a problem relevant for the future profession of these students, being it in medical, engineering or other problem-oriented disciplines. Such experiential learning (the term is introduced by Kolb 1984) in a (simulated) context enables students to gain expert experience during their studies.

## References


Blom, A.F. (1982) *Het KB2–Project. Verslag van een Onderwijsexperiment* (The KB2–project. Report of an educational experiment). (Wageningen, the Netherlands: Wageningen Agricultural University).

Ertsen, M.W. (1999) "Designing as a continuous learning process. Attempts to educate a learning designer." 4th European Forum for Continuing Engineering Education, Trondheim.

Ertsen, M.W. and Heijer, T. (1998) "A systems approach to sustainability in water issues." 1st Inter-regional Conference on Environment-water: Innovative Issues in Irrigation and Drainage, Lisbon.

Herder, P. (1999) "Process Design in a Changing Environment." PhD thesis, Delft University of Technology.

Jong, de M. (1999) "Institutional Transplantation." PhD thesis, Delft University of Technology.

Keulen, H. van (1998) "Ontwerpen van ontwerponderwijs" (Designing design education). *Onderzoek van Onderwijs* (Research on Education).

Kolb, D.A. (1984) *Experiential Learning. Experience as the Source of Learning and Development* (New Jersey: Prentice-Hall).

Krick, E.V. (1969) *An Introduction to Engineering and Engineering Design* (New York: John Wiley and Sons).

Mast, C. van der, Sodoyer, B. and Kunst, S. "Multimedia-onderwijs aan de Technische Universiteit Delft" (Education in multimedia at Delft University of Technology). *Tijdschrift voor Informatica-onderwijs* (Periodical for education in informatics) at http://is.twi.tudelft.nl/~charles/in457/Tinfon/tinfon.html

Meijer, A.W.M. (1998) "Wat maakt een ingenieur?" (What constitutes an engineer?), Inaugural speech, Delft University of Technology.

Meijers, T. (1990) *The Interaction Between Users and Designer in the Design Process of Village Irrigation Systems on the Island of Ile a Morphil, in North Senegal. Design for Sustainable Farmer-managed Irrigation Schemes in Sub-Saharan Africa* (Wageningen, the Netherlands: Wageningen Agricultural University).

Norman, G.R. (1988) "Problem-solving skills, solving problems and problem-based learning." *Medical Education*, 22, 279–86.

Payne, A.C., Chelsom, J.V. and Reavill, L.R.P. (1996) *Management for Engineers* (Chichester: John Wiley and Sons)

Ploeg, J.D. van der (1991) *Landbouw als Mensenwerk. Arbeid en Technology in de Agrarische Ontwikkeling* (Agriculture as work of man. Labor and technology in agrarian development), (Muiderberg: Coutinho).

Scheer, S. (1996) "Communication Between Irrigation Engineers and Farmers." PhD thesis, Wageningen Agricultural University.

# 14 Making connections

## Design and the social sciences

*Desmond Rochfort*

### Introduction

In the Canadian daily, the *Globe and Mail*, dated 26 July 1999, the front page of the section on Careers and Managing carried an article entitled, "Firms Seek Guidance From Anthropology." Below, the bi-line read: "Specialists in the academic discipline—experts at observing, documenting and analyzing human behavior—are attracting a following among corporations eager to know what makes their workers and customers tick." The article, focused on the work of an anthropologist; "These days," it read, "(Ms. Squire) is far more likely to find herself in a board room . . . As a consultant in Palo Alto in California—the heart of Silicon valley—she uses her training in the study of human behavior and culture to develop new products. . . ."

This example of interconnective thinking between two apparently quite separate domains of activity and enquiry, in this case, between business enterprise and a social science discipline, clearly has important implications for design practice, where the connective links to other practices and modes of enquiry are in some cases quite explicit.

Though the notion of interconnectivity, and interdisciplinarity is inherent in any contemporary considerations of design thinking and practice, nevertheless we perhaps ought to consider for a moment the need for breaking some connections. In particular the connection that is made on a daily basis, in public discourse, between "art" and "design."

To propose that design and art are for all useful purposes not connected with each other is to move into intellectually contentious and dangerous territory. Historical, as well as institutional convention, would certainly lead anyone to believe that theirs was an intimate and singular relationship. The ease and habit with which we couple the words "art" and "design" together in our conversation, the frequency with which we put them together under the same educational roof, serves to reinforce, even if subconsciously, that theirs is a natural, symbiotic relationship, one that would be unthinkable to configure in any other way than that which historical convention has up to now presented us with.

However, the relationship and connection that does exist between design and art, as it is presented today, is in many senses false, contradictory, and often unequal, and is becoming more tenuous with each passing year. Certainly the accepted, if not imposed relationship of design with art, has often had a diverting if not damaging influence on the way design education is conceived and has been developed and carried out in many institutions.

In part, history has not helped us. Indeed it has dealt us a confusing narrative of the way in which different disciplines within the creative realm of the visual have been configured. In ancient Greek culture the practice of sculpture, painting and architecture were all regarded as architecture. In the Italian Renaissance of the fifteenth century, painters and sculptors began to separate themselves off from the definition of being judged as mere craftsmen. Their increasing interest in scientific and technical issues, such as for example the laws of perspective meant, as Frederick Antal observed, that their technical skill was increasingly put "more and more to the service of scientific innovation based on theoretical knowledge: for it was by establishing a theoretical and scientific foundation for itself that art could obtain greater social recognition, could free itself from the crafts, and so those who practiced it could rise from the condition of artisans" (Antal 1979: 376).

The great leap in cultural and social significance that the art of painting and sculpture made during this time from that of artisan and craft to sit alongside the great liberal arts, in part can explain why it is that design as a discipline today still has difficulty, within the academy, of defining and creating its separate existence from art, understanding the rationale of its own history and unsure of its methodologies. Indeed the emergence of the academies of art in the sixteenth and seventeenth centuries created a framework of practice and thinking that when eventually design did emerge from art's shadow during the Industrial Revolution, it did so almost as a step-child of the fine arts, particularly in the United Kingdom, with all the consequent contradictions, anomalies and uncertainties which we see still see today in approaches to design education. All this despite the impact of the Industrial Revolution, the rise of mass production and mass consumption, and the consumer market, in which design rather than art has played a pivotal part.

Before embarking on some thoughts on the challenges and opportunities that the academy faces in addressing some fundamental issues confronting design education, it is worth perhaps rehearsing some of the profound differences that exist between design and art. For I would argue, that it is only when we really understand the differences can we begin to construct a design education that responds to and reflects the intrinsic nature of the design discipline and the design process, and so finally free it from the notion that design is really only a sub-discipline of art.

Although differences have always existed between the notion of design and what became known, from the fifteenth century onwards as the fine arts, historically the emergence of design as a fully recognized and distinct category of practice took place in the nineteenth century during the Industrial Revolution. Then the introduction of new industrial technologies of production and manufacture created the basis for mass production, and the development of the mass consumer market. This led to the rapid development of urbanization, which in turn fueled an increasing need for artisans capable of creating communications and products for the new urban mass markets and their growing middle class. In today's infinitely more complex and sophisticated societies the practices of design with its various sub-disciplines such as Visual Communications Design, Industrial Design, Interface Design, Interactive Design, etc., are even more distinct from the fine arts, in terms of applications, methodologies, to say nothing of the differences in terms of social and economic impact.

Of course the disciplines of design and the fine arts share the common root of being grounded initially in visual language. And of course the relationship to each other as well as the histories of the different manifestations of creative visual activity existing in these broad constructs, is more complex, textured, and qualified than this schematic and brief narrative and arguments might imply. Nevertheless, it is worth repeating that the last two hundred years have seen differences between design and art emerge and diverge in a way that demands that we now recognize that both now exist in quite separate disciplinary domains, with profoundly different discourses and intentions, with significant implications for how we approach the education of designers.

So what are the differences? Essentially the differences between the two domains lie in the fact that the fine arts are for the most part based around single "artist" or single "maker" centered practices. In these "aesthetics" the imperative of "self-expression" and the acquisition of a "personal practice" converge as primary attributes of this creative cultural paradigm. On the other hand, excepting such areas as haute couture fashion design, where self-expression and the personal "stamp" of the designer predominate, most areas of design today would not regard these attributes as prerequisite guiding principles.

Although design obviously engages with aesthetic considerations, its primary *raison d'être* is not located around the maker of the design but rather in its "end-user." This results in a practice that involves multi- and interdisciplinary tasking, applications, methodologies and knowledge to identify and solve problems and tasks that arise out of the ebb and flow of demands and needs in society.

In many respects design now occupies a "third area" between the humanities and science. The British designer and educator Nigel Cross in a response to a major report commissioned by the British Secretary of State for Education more than twenty years ago, entitled Design in General Education, wrote: "The sciences value objectivity, rationality, neutrality, and a concern for the 'truth.' . . . The humanities value subjectivity, imagination, commitment and a concern for 'justice.' The designerly way of knowing involves a combination of knowledge and skills from both science and the humanities. . . . Design has its own distinct things to know, ways of knowing and ways of finding out about them'(Cross 1979: 221–2).

Although visual communication design is different from that of automotive design, in terms of scale, and the ultimate intended use, they nevertheless share many of the same methodological processes for both identifying and solving problems as other design practices. Although their processes do not involve the same degree of specificity as the scientific method, they are nevertheless infinitely more explicable and analytical in idea development than the speculative, self-expressive approaches of fine art practice. Their methodological process begins, as Meredith Davis has observed, "with the identification of a problem, involves the research and the ranking of competing priorities, that often appear to be in competition with one another, tests the variability of multiple solutions through prototypes, and ends with the evaluation of objects against a socially mediated set of performance criteria" (Davis 1998).

In almost all cases, from the initial identification of a problem through to the proposed design solution of the original identified problem, the design process is essentially social, multi- and interdisciplinary, rather than individual in character, engaging and interfacing with its users or audiences, who determine its outcome and effectiveness by the way and by the extent in which they make use of it.

Today, the solution to a design problem may reside in the knowledge and practice of a whole range of other disciplines. The design and styling of an automobile, for example, will inevitably involve the need for and application of scientific and engineering data, in addition to arrays of consumer data, market segmentation analysis and so on. Likewise, designing and developing a communication structure of easily accessible and effective signage and information for international airports, in which millions of people of different nationalities speak numerous languages (perhaps one of the most challenging of information and visual communication design tasks), will almost certainly incorporate methodologies and information derived from psychology and other social science disciplines. As a result of the very nature of both the design process and its application as a practice, design is moving from being simply designing *for* users to being one of designing *with* users. Such differences now glaringly separate design from the fine arts, both at the professional level and importantly and necessarily at the educational level as well.

So what are the implications for design education? What should design education be doing to reflect what clearly are changing needs for a differentiated set of skills and intellectual competencies in the new designers of the future? In a recent briefing paper of a report compiled by the American Institute of Graphic Arts and the National Association of Schools of Art and Design entitled "Technology Thresholds in Graphic Design programs" (Davis, no date) the authors wrote:

> It is virtually impossible to practice graphic design today by using only traditional hand processes, such as physical paste-up of mechanicals. Photo-type setting and retouching have been transformed into electronic output and pre-press services and the once separate functions of graphic design and production have collapsed into a single effort now often under the control of the designer. . . . It is now assumed by employers that all students entering the field from undergraduate study will have the ability to: author text in word processing programs, draw graphic images on the computer, manipulate photographs digitally, produce digital page layout, understand related issues to output and electronic pre-press, at least in terms of file preparation; Choose appropriate technological resources for specific design tasks.
>
> (AIGA/NASAD)

And over in the next page of the briefing report there is the observation that the explosion of new media has changed the work in design offices from print-based projects to include electronic communication and concludes that most undergraduate students will also be significantly more employable if they possess a rudimentary ability to: work in time-based media, design motion typography, design information architecture, interfaces and narratives for the internet, design time-based media in cross-disciplinary projects.

Of course one cannot disagree with any of these recommendations. Far from it. How could anyone confronting the technological changes taking place deny the necessity of competencies in these areas for the young designer freshly minted from their degree program. And yet is there not something missing? If it is true that the design process is essentially a social, multi- and interdisciplinary activity, engaging and interfacing with its users or audiences, who determine its outcome and effectiveness by the way and by the extent in which they make use of it, then clearly, even at the undergraduate level, competencies focused exclusively on technological applications are not sufficient.

Today, design degree programs, even at the undergraduate degree level must, as part of their central aims and objectives, equip their graduates with a knowledge and an intellectual and theoretical grasp of those bodies of knowledge and methodologies that exist outside of design but which design increasingly calls upon and relies on to help develop and conceptualize solutions to identified problems. Not to do this reduces the educational process merely to the "grunt-work" of applications training, which in the end serves little useful purpose, and greatly hinders the potential as well as the capabilities of the designer to bring about effective solutions.

This line of argument predictably leads to very specific questions of curriculum content and structure, and the relative distribution of time allotted to different parts of an overall curriculum in design. At this present time this is less important than the need to recognize that design education is entering a time of transformational change. It is doing so precisely because design practice is changing. And design practice is changing because it is becoming increasingly recognized that design, in all of its various manifestations, some of which are visible and some of which are not, has a much more significant and influencing role to play in not only how we see our socio-economic environment but also what we make of it.

If design is not art, then it is important to stress that it is also not science, social science, or engineering. Rather, as Charles Owen, Professor of Design at the Institute of Design at IIT in Chicago has observed, design is integrative (Owen 1989: 3). But what does this actually mean in practice? Essentially it means that the design process, and with this effective design, is a central component of a linked continuum of thinking and practice, a kind of seamless interconnected web of knowledges, practices and methodologies being operated in synthesis towards a common goal or objective—at least that is the ideal of a design process.

From the point of identifying a problem, through to the conceptualization of the solution to the problem, and on to its deployment as a user-oriented outcome, the design process involves an intellectual and conceptual journey, a narrative of linkages, associations and dependencies. Through this journey the designer will encounter, work with or need to use different knowledges, different methodologies, whether of analysis or planning, techniques, skills, applications and languages. Some of these will be ones we conventionally connect with the work of a designer, in that they will be grounded in visual concepts, applications and language, but many will not. Break these linkages, or fail to build them into the design process narrative and an effective outcome will become wanting.

In one sense there is no better example of the need for integrative thinking and practice than in the development of a computer software program. A computer program begins with science, with math, algorithms and calculations, programming protocols, and the application of coding language as obscure and as opaque as the popular imagination could conceive. During the journey to its completion the science and the math are manipulated and fashioned so that their invisibility can be rendered visible and useable, to realize a specific objective, whether as a tool, or as an entertainment, or even as a medium. As users we seldom ever need to "see" or even "use" the math, the calculations, the coding language. Because from the moment the mathematics of coding language is conceived to become or is transformed into a visual sequential or interactive interface—and practically all computer software programs are conceived as visual interfaces—the math and the science disappear and what is revealed instead becomes a matter of design, and the imperative of the design concept takes over. The design of course will have an end-user, who may either be a consumer using the program for limited and quite specific purposes or someone or some organization that requires it to perform much more complicated and sophisticated tasks. The problem with so much computing software therefore is not so much the science but the design. And the problem with the computer software design is that it necessarily engages or should call upon high levels of multiple competencies, in coding language, in creative and graphic concept development, interactive and interface design, and cognitive factors relating to the interface and the interactive environment being created, to name just some. All too often the key component parts in the program's design narrative are not working and talking with each other in an integrative sense, or if they are, they are often not doing so with an understanding of each other's language or more importantly *modus operandi*. This critique is not to suggest that computer scientists, programmers, and coders, should suddenly all turn themselves into visual communication designers. Nor conversely is it suggesting that designers should transform themselves into computing scientists, become expert programmers and coders, train to be cognitive psychologists, as well as acquire terminal degrees in business management, accounting and marketing. Rather what it is suggesting is that there is most definitely a need for the development of an interdisciplinary, multidisciplinary framework of competencies within design. If design is a multitasking operation, then the designer needs to develop some level of multitasking competencies if he or she is to bring about the eventual deployment of a successful and effective "solution" to a problem identified and then defined.

How one constructs the dynamics of an integrative multitasking thinking and practice into design education in a meaningful and effective way is not without its challenges, not least because surprisingly all too often design is still viewed as maker-centered and not user-centered, a creative variant of art but with utilitarian overtones. As such, therefore, many design programs, while they may pay lip service to it, seldom provide a suitable educational framework that instills in the education of the young designer an intellectual and theoretical grasp of the integrative multitasking imperative of the design process. Nor, as a consequence, have they constructed a realistic framework for the development of multi- and interdisciplinary competencies, in which the integrative imperative of design can be learned through practice within their program.

Of course, as a prerequisite it is absolutely essential for design education to create a foundational framework of learning that equips a design student with the ability to reach across disciplines, to bring in information, to extract ideas, to think critically and to make connections. This calls for a skillful blending of arts and humanities, sciences and technology, as a core series of intellectual and knowledge-based building blocks around which design education is threaded. And many design programs and schools know this perfectly well and some practice it very well. But it seems that there must be opportunity, even within the framework of undergraduate design education, to begin to build onto the studio based aspect of design education, interdisciplinary competencies engaging those disciplines beyond the studio with which the design graduate will have to integrate in professional practice.

Five years ago the Department of Art and Design at the University of Alberta embarked on an ambitious project to develop precisely this kind of educational framework. As part of its Bachelor of Design degree, the department created, in collaboration with various faculties at the university, a series of intensive pathways of study in disciplines that it considered to be relevant for a student studying either visual communication design or industrial design—the two streams within its design program.

The alternative pathways of study open to design students to choose from were in:

- Computing Science, which is taught out of the department of Computing Science.
- Engineering taught out of the faculty of Engineering.
- Business and Marketing taught by the faculty of Business.
- And from the Fall of 1999, the introduction of a further alternative pathway of study, in the social sciences of Anthropology, Psychology and Sociology, taught of course by faculty from these departments.

Each of these pathways of study—a student can only choose one due to the intense nature of the pathway study—runs alongside and parallel to the student's studio-based design courses, and does so throughout all four years of the degree program. Depending on the pathway chosen the number of credits in one of these pathways can run between a quarter to nearly one half of the total number of credits in the degree as a whole. But whatever the intensity of the pathway, studio design course work is not sacrificed.

The aim of the pathways is not to pretend to graduate qualified computing scientists, engineers, business managers, sociologists, anthropologists or psychologists. Rather they were conceived to provide a design student the opportunity to develop some core competency and understanding in disciplines of particular relevance to the areas of design they were concentrating their studies in.

This initiative, taken five years ago, was one small step towards conceiving of a different framework for design education. At this present moment the character of this initiative is what could be described as "disciplinary parallelism." In other words, a course of study in a subject is run alongside another course of study in another subject. Whilst this a small step, it is a very important one, for aside from developing some understanding and competency in an associated discipline, it crucially lays the ground for the development of what I would consider to be the next critical step, and that is "interdisciplinary integration." This will occur when the course of study in design and the course of study in the pathway discipline are pedagogically synthesized into a single integrated course of study.

The implications of this development are enormous in their potential for both design education and as importantly for design practice on a professional basis. And in addition it would lay the ground work for some significant developments in this direction in graduate programming and research. But I do not underestimate the institutional and pedagogical challenge of achieving this degree of integration. There is much work to be done pedagogically in this respect, in the building of alliances and the clarifying of the intellectual and creative aims and objectives and ultimate goals of design education at the undergraduate level within the context of change that confronts design. Each educational institution is different, and each will find its own way that is appropriate and suitable for its circumstance. But whatever the circumstance and whatever the way, the interdisciplinary-integrative imperative can no longer be avoided.

## References


Antal, F. (1979) *Florentine Painting and its Social Background* (New York: Harper and Row).

Cross, N. (1979) *Design in General Education* (London: Royal College of Art).

Davis, M. (1998) "Making a case for design-based learning." *Arts Education Policy Review*, 100 (2), November/December.

Davis, M. (ed.) (no date) *Technology Thresholds for Graphic Design Programs*. A paper prepared by a joint AIGA/NASAD task force led by Meredith Davis, North Carolina State University. Published by the AIGA at URL: http://www.aiga.org/Lindex.cfm?CategoryID=34

Owen, C. (1989) "Design education in the information age." *Design Processes Newsletter*, 3, 3.

# 15 Appearance, form, and the retrieval of prior knowledge[1]

*Peter Dixon and*
*Tenaha O'Reilly*

## Introduction

In our experience, adults almost never learn completely new procedures. Instead, when confronted with a new task, device, or situation, what people almost always do is to adapt some old, previously learned procedure to the new task. For example, when learning to use a new digital camera, one draws on knowledge about how to operate film cameras; when learning to use a video disk player, one uses knowledge about how to operate a VCR; and so on.

Occasionally, there is no body of comparable or related procedural knowledge to fall back on, but in situations like these, people generally perform very poorly and have nowhere near the facility that they have in learning to use a new printer or different model telephone. For example, in learning to use a global positioning device, a user may be completely at sea, even though on some level the task may be no more complicated than programming a VCR. The moral that we draw from such observations is that learning new procedures is generally successful because the "new" procedures are actually very similar to old procedures and do not require learning something entirely new from scratch. On this view, most procedural learning in the world is critically dependent on the recruitment of prior knowledge of similar tasks. The corollary is that if one wants to understand how people learn new procedures and to use new devices, one needs to understand the nature of procedural similarity and its effects on learning and performance. In the present article, we attempt to move in the direction of that understanding. In particular, we present an overview of some of the evidence and insights that have resulted from several years of laboratory research on the problem of how prior knowledge of similar devices affects the learning of new operating procedures. We believe this research has some important implications for design, and these will be mentioned in the conclusion.

## Plans and procedures

We begin by describing a theoretical framework and terminology for describing similarity between procedures. This classification scheme is process-based; we believe that a good metric of similarity between two procedures should be based on how similar the mental processes and representations are in carrying out those procedures. In turn, this means that a prerequisite for understanding similarity is an understanding of what it is that people do when carrying out a task. Such understanding is available, at least in broad outline: tasks are performed by generating mental plans for the task and then carrying them out (Dixon 1982). Further, there is good reason to believe that these mental plans have a hierarchical structure (e.g. Miller *et al.* 1960) with at least three qualitatively different levels. We use the terms sub-goals, steps, and motor-control schemas to refer to these levels; below, we discuss each in turn.

The top level of a mental plan consists of a sequence of sub-goals. These are high-level goals corresponding to states or modes of a device or interface. For example, the sub-goals involved in using a digital wristwatch as a timer might consist of first going to timer mode, setting the time, and then running the timer; performing the task as a whole consists of carrying out each of these sub-tasks one after the other. At the next level, each sub-goal can be broken down into a series of steps; this step level usually corresponds to the sequence of sentences that might be generated in a verbal description of how to accomplish a sub-goal. For example, on a particular watch, getting to the timer mode might involve pressing the "mode" button two times until the word "timer" appears on the watch display, and this is exactly how a set of directions for the task might be phrased. Finally, at the bottom level, each step can be expressed as a motor-control schema that identifies the physical actions and constraints necessary to accomplish a particular step. So, for example, pressing the mode button on a given watch might involve holding the watch with one's index and thumb on opposite sides with the thumb over the "mode" button, and squeezing momentarily. Although the motor control details needed to carry out a given step are clearly essential for task execution, they are rarely expressed in verbal directions.

## Investigating effects of similarity

The reason these notions of planning are important is that similarity between plans can occur at any of these levels. We use the terms "functional similarity" to refer to procedures that involve the same sequence of sub-goals; "procedural similarity" to refer to situations in which the same steps are used to achieve a given sub-goal, and "interface similarity" for situations in which the same motor-control schema is used to carry out a given step. Crucially, each kind of similarity has potentially different effects on procedural learning. However, these different effects can be difficult to discern in the real world because different types of similarity tend to go together: by and large, devices that do the same thing are functionally similar, use similar procedures, and have similar interfaces. In particular, forms of similarity have the same hierarchical arrangement as the procedural plans do, so that devices that have same interface typically have same operating procedure, and having the same operating procedure generally means working on same set of task sub-goals. For example, starting a car is pretty much the same regardless of its make and model: the interface is the same, the sequence of steps that need to be carried out is the same, and the same set of intermediate states are involved.

In order to tease apart these different forms of similarity, we have people learn precisely controlled, artificial tasks in a laboratory setting. Further, in order to control what people know that is relevant to learning a novel task, we use transfer paradigm. The general approach is illustrated in Table 15.1. Subjects in a transfer experiment are taught two tasks. The first is the training task and the second is the transfer task. Generally, subjects will be somewhat better at learning the second task than the first; this improvement is transfer. For technical reasons, we measure transfer as the logarithm of the number of errors made in the first task divided by the number of errors on the second. Thus, large values of transfer mean that a great deal of what was learned on the first task carried over and could be used on the second task; small values near zero mean that the acquisition of the second task was little better than the acquisition of the first. The amount of transfer is examined as a function of the relationship between the two tasks. In a similar-task condition the training and transfer tasks are similar in some specific respect, while in a dissimilar-task condition that element of similarity is removed. Crucially, the very same tasks are used in both conditions except for the critical manipulation of similarity. Thus, the difference between the amount of transfer in the similar- and dissimilar-tasks conditions provides an unequivocal measure of the effect of that particular form of similarity.

To anticipate, the results of a variety of experiments suggest that there are distinct effects of each type of similarity, and, as near as we can tell, these effects are independent of one another. What this suggests, in other words, is that each form of similarity has an effect on a different mechanism involved in learning. Below, we provide an example of some of these

*Table 15.1* Investigating similarity with a transfer paradigm

| | *Training task* | *Transfer task* |
|---|---|---|
| Similar condition | Task 1 | Similar task 2 |
| Dissimilar condition | Task 1 | Dissimilar task 2 |

conditions and the corresponding effects on transfer. A learning mechanism that might be involved is discussed in each case.

## Effects of functional similarity

Perhaps the least successful of the manipulations we have investigated concerns functional similarity. In many transfer experiments like those outlined above, we have failed to find strong effects of functional similarity. However, one context in which such effects are found involves the operation of completely novel devices in an unfamiliar domain (O'Reilly and Dixon 1999a). This, of course, is easily accomplished in the laboratory by creating a fictitious task and device that no one has heard of before. The task in this case involved operating an automated remote weather station that collects and records meteorological data from a variety of sensors. Figure 15.1 depicts an example of one of the interfaces subjects used. Subjects were given a task involving a series of startup and maintenance procedures. For example, as part of their task they were asked to perform an operation called "data relocation" that was accomplished by pressing buttons labeled "transfer," "system," "externals," and then "retain." After learning these procedures, subjects were given as a transfer task a different version of the weather machine (Figure 15.2) with analogous procedures and steps. For example, they might be asked to do the same data relocation

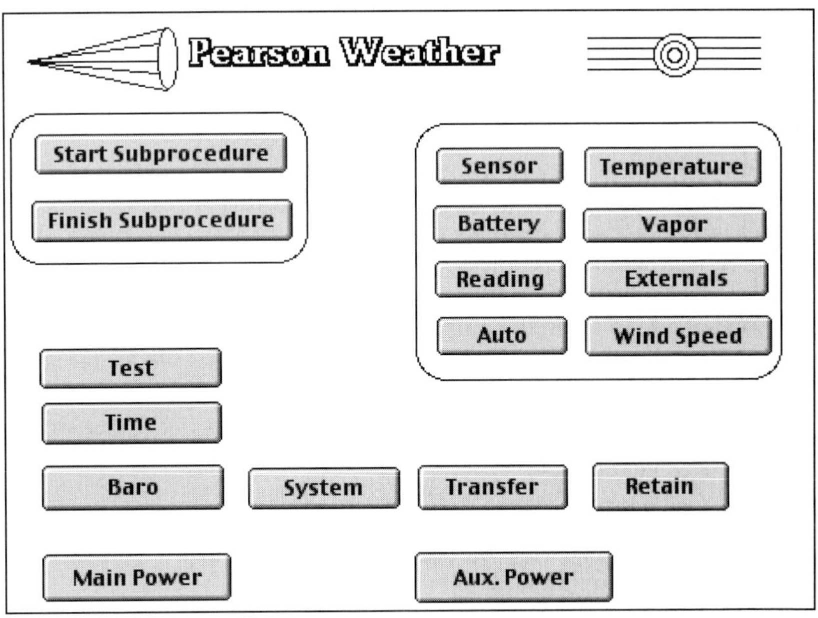


*Figure 15.1* Examples of the interface display used in the training task for the investigation of functional similarity.

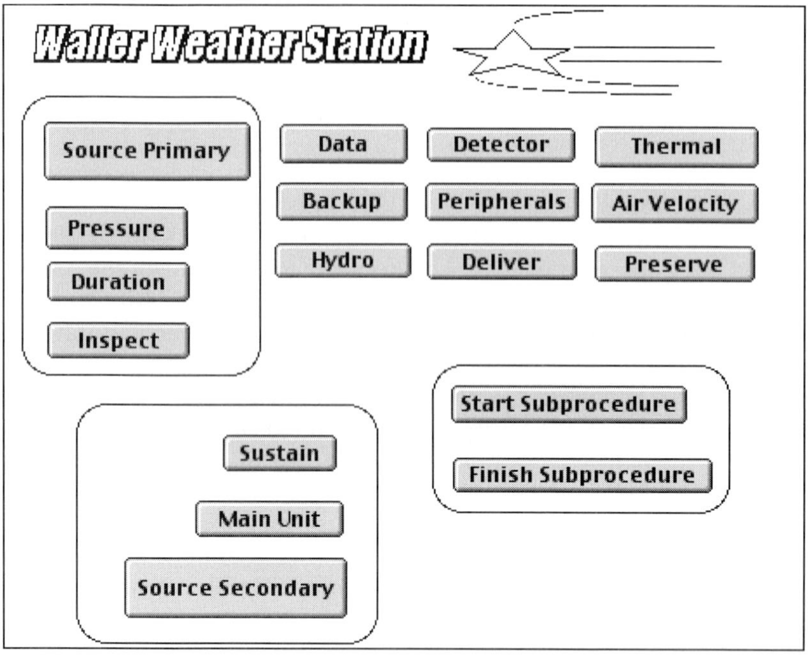


*Figure 15.2* Example of the interface display used in the transfer task for the investigation of functional similarity.

operation by pressing "deliver," "main unit," "peripherals," and "preserve." In this case, the operation would be accomplished using the same steps with buttons that have different but roughly synonymous labels. The two devices in this condition were thus functionally similar.

However, other subjects were given the same transfer device with precisely the same sequence of steps but with a different description and explanation for each of the sub-procedures. For example, subjects might be asked to perform an operation called "system cleaning" with same sequence of button presses, "deliver," "main unit," "peripherals," and "preserve." In this case, there are different sub-goals for the task, presumably involving different internal states of the device. Thus, the two tasks were functionally dissimilar.

The results (Figure 15.3) indicate that substantially more transfer was obtained when same sub-procedures were used in the training and transfer tasks, even though the steps were exactly the same in both conditions. Our interpretation is that being familiar with a task and a device gives one some understanding of how the device works. In turn, this knowledge allows one to infer or anticipate what the appropriate sequence of sub-goals should be for a novel device that works in the same way. Without this familiarity, extra effort is required to acquire a grasp of the new sub-goals for the novel task, and less transfer results.

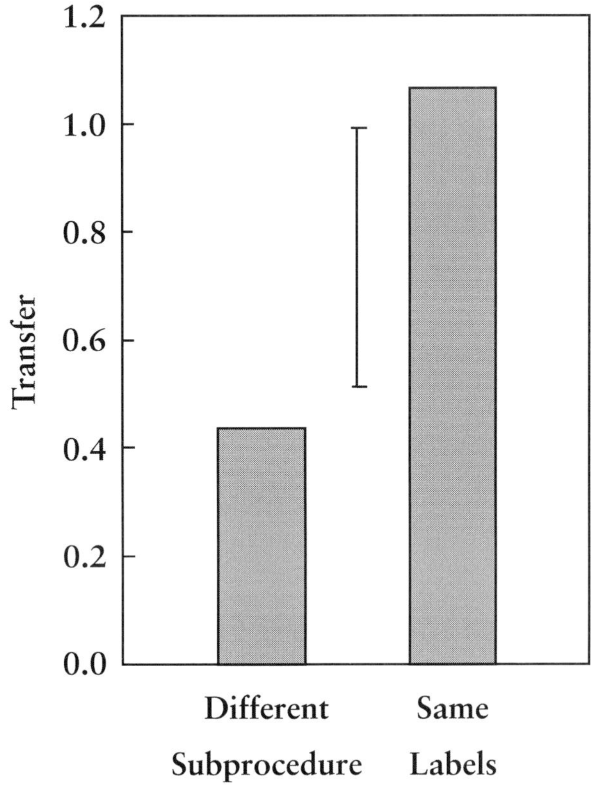


*Figure 15.3* The effect of functional similarity on transfer. The error bar indicates half the size of the 95 percent confidence interval for the difference.

As noted above, this result has proven to be the exception rather than the rule in our research; often, we find minimal or non-existent effects of functional similarity (e.g. Dixon and Gabrys 1991). We think that this is because most devices that have been used, although fictitious, are from relatively familiar domains with relatively intuitive goal hierarchies. For example, we have devices such as video disk players, burglar alarm systems, and airplane control panels. Because the operation of these devices is relatively natural and intuitive, the sub-goals are easy to learn even in the training task. Consequently, there is little additional benefit from previous exposure to a functionally similar device. In fact, this is generally true of devices encountered in the world: typically, when we learn new devices and procedures, the procedures are likely to be comprised of familiar, well-understood tasks. For example, when you get a new phone, you expect to be able to make and answer calls, even if the buttons and indicators are new; when you get a new printer for your computer, you expect to be able to print documents, even if paper loading and power up procedures are not familiar; and so on. What this means is that an appropriate sequence of subgoals will be easy to acquire and there is likely to be little additional benefit of a device being functionally similar to other devices.

## Procedural similarity

In contrast to functional similarity, we have found a host of manipulations of procedural similarity that produce large effects on transfer. Some of these manipulations include varying the order or nature of the steps needed to accomplish a sub-goal, the precise nature of button labels, and variations in formal or "syntactic" steps like pressing a "mode" or "set" button. In one experiment, we simply changed the order of the steps in a sub-procedure from training to transfer (Dixon *et al.* 1997).

Subjects in the study were given a control panel and told that it operated an airplane; their task was to perform a sequence of sub-procedures for powering up the electrical system, starting the engines, setting up the communications system, and doing a communications test. In one instance, the sub-procedure of starting the engines consisted of pressing buttons labeled "Engine 1," "Engine 2," and "Ignition." In the subsequent transfer task, some subjects had comparable steps in an homologous order, while others had the steps in a different order. For example, in the similar condition, the sub-procedure for starting the engines required buttons labeled "Prop 1," "Prop 2," and "Starter" be pressed; in the dissimilar condition, the "Starter" button would need to be pressed first followed by "Prop 1" and "Prop 2."

The results (Figure 15.4) show what one would expect: there was a substantial benefit from having previously learned the steps in same order. Our interpretation is that this effect is not at the level of sub-goals or internal states of the device and does not involve reasoning about the device and how it works. Similarly, this effect cannot be at the level of physical actions since the buttons to be pressed were always in different locations and labeled differently. Instead, we argue that this effect is memory based: during training, subjects learn a sequence of steps for accomplishing a particular sub-goal, and then retrieve that sequence during the transfer task. If the retrieved information matches the procedure in the new task, learning that task is much easier. We refer to the retrieved sequence of steps as a routine; we hypothesize that such a routine is a simple verbal representation, akin to a list of steps. Further, we argue that routines of this sort are the basis of a great deal of the transfer that occurs at the level of procedure steps.

## Interface similarity

Our work on interface similarity is in its early stages, and we have only investigated a few manipulations so far. One simple manipulation involved

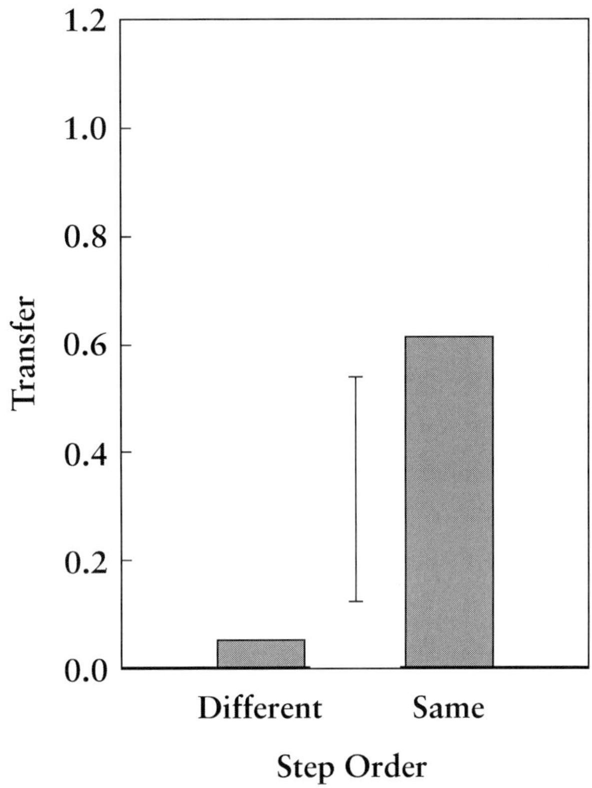


*Figure 15.4* The effect of procedural similarity on transfer. The error bar indicates half the size of the 95 percent confidence interval for the difference.

varying the positions of buttons on the control panel (O'Reilly and Dixon 1999b). This involved giving subjects two devices with the same buttons, labels, and steps; the only aspect of the devices that was manipulated was the spatial arrangement of the buttons so that the physical actions needed to carry out a step could be either the same or different. The results (Figure 15.5) demonstrated more transfer with same layout from training to transfer than with a different layout.

A critical aspect in this study was that the same effect of button layout was observed regardless of whether the labels on the buttons were same or different. This means that the effect of interface similarity cannot be mediated by a verbal representation of the steps of the procedure, and must be due to a different mechanism than the one implicated in the effect of step order. We believe that the effect of interface similarity is at the level of action plans and is related to the interface affordances. We understand affordances in this context to be the match between what the user can do and what he or she intends to do. For example, the subject may want to press, at a given point in the procedure, the wind speed button. When the

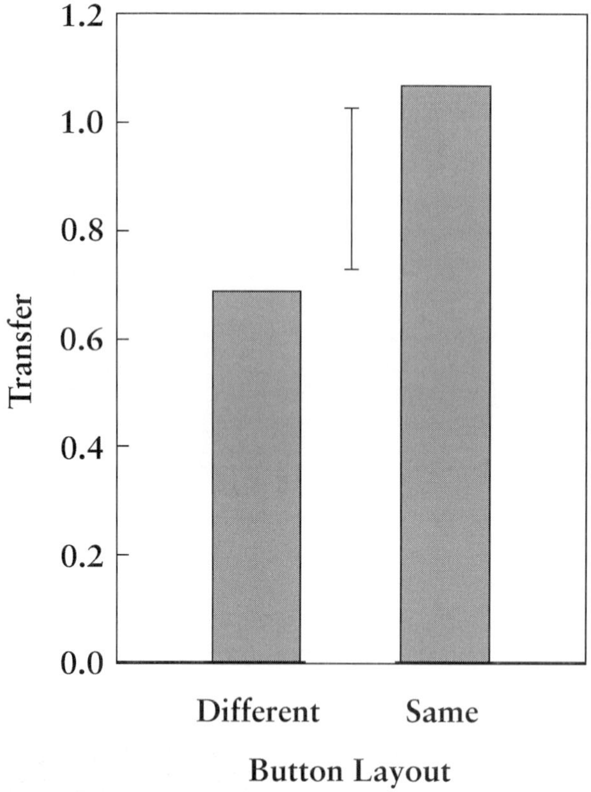


*Figure 15.5* The effect of interface similarity on transfer. The error bar indicates half the size of the 95 percent confidence interval for the difference.

layout of the buttons is similar, subjects can use the same affordance learned for that step in the training task, and move to same location to press the button. But when the layout is different, new affordances would have to be generated, and transfer would be reduced. It would be difficult to attribute this effect simply to a facility in generating various actions since there is often a button at a given location even if the layout has been rearranged. Instead, what we believe is crucial is that there is a match between the intention to press a particular button and the required action. More generally, we suspect that the notion of affordances, in a perhaps more sophisticated form, can explain a variety of the effects of interface similarity.

## Similarity and learning mechanisms

In some ways, the results summarized here may seem fairly obvious: essentially, we have shown that new procedures are easier to learn when they are similar to previously known procedures. In other words, what we are describing are merely demonstration experiments for various aspects of learning that might be expected on fairly intuitive grounds. On the other hand, it is important to note that there are many manipulations of similarity that have relatively little effect on learning. For example, there is little effect of manipulating the overall appearance of a device by varying the graphics, fonts, and logos; also, simply having the same controls on a device has no effect on learning if those controls are used for different purposes; or, having the same button names helps very little if the steps are performed in a different order; and so on. In other words, only some kinds of similarity are important in certain contexts. As argued at the outset, similarity needs to be understood in terms of the process of carrying out the procedure, and it is only in that context that these effects become relatively predictable.

Our main point, however, is not that similarity helps people acquire procedures, but rather that there are different kinds of similarity, and the kind of prior knowledge that gets recruited by each is different. In particular, we argue that there are at least three different kinds of representations and mechanisms involved in transfer, with at least one corresponding to each form of similarity. These are summarized in Table 15.2. First, functional similarity between tasks allows one to use knowledge concerning intermediate states of the device or interface and allows one to reason or make inferences about those states. Second, procedural similarity between tasks allows one to retrieve and use routines, that is, sequences of steps that have to be performed in a particular context to achieve a given goal. And third, interface similarity allows one to generate appropriate affordances, in which the intention and the design of the interface combine to afford appropriate actions rather than inappropriate ones.

Further, these can be quite separate, independent mechanisms, and one can observe a benefit of one kind of similarity even if the tasks and devices are quite dissimilar in other ways. A valuable (but perhaps dated) example involves learning to use a word-processing program on a computer when one is only familiar with word-processing using a typewriter. Our analysis is that the functional similarity between using a typewriter and using word-processing software is very low and whatever relationship exists is likely to generate negative transfer. For example, an informal observation is that in this situation it was fairly common for a skilled typist to insert a carriage return at end of each line when first learning to use a word-processing

*Table 15.2* Planning levels, forms of similarity, and learning mechanisms

| *Planning level* | *Form of similarity* | *Learning mechanism* |
|---|---|---|
| Subgoals | Functional | Reasoning and inference |
| Steps | Procedural | Routines in memory |
| Motor control schemas | Interface | Generation of affordances |

system. The procedural similarity between these two tasks is also low, and the majority of steps performed on computer have no analog on a typewriter. However, the interface similarity is high: both use a keyboard with a virtually identical layout to type words. We suggest that this similar interface allows the learner to generate common affordances for entering words and sentences. As a consequence, typing skill transfers almost completely and allows one to be much more effective in computer word-processing, even though an entirely new sequence of goals and steps has to be acquired.

## Implications for design

We believe that our analysis and research has some implications for design. Although some aspects of devices and their associated operating procedures are determined by relatively immutable engineering considerations, there are other aspects that are more arbitrary and often end up being a function of someone's design conception. For example, our suspicion is that the layout of the buttons on a VCR remote is relatively unconstrained; similarly, there is likely to be a wide range of different ways in which the steps for operating a microwave oven could be arranged. Our view is that in designing features such as these, one should not minimize the impact of learnability, and, in the design stage, one should carefully consider how easy it is going to be for someone to acquire the operating procedures for the device. Although people can learn to do a wide range of difficult tasks if they have to, there are many situations in which people will simply refrain from using a device if the procedures are difficult to figure out; this is usually not a desirable design outcome.

We argue that a crucial determinant of learnability is the similarity of the device and operating procedures to known, familiar devices and procedures. What this implies is that, other things being equal, devices should be designed so that they are similar to things people already know how to use. However, this has to be done in an intelligent way, since the form of the similarity determines what prior knowledge will be used, and consequently what the impact will be on learning and usability. The present research provides the basis for a variety of recommendations on how this might be done. For example, designing a device with a familiar interface would allow the user to immediately appreciate how particular steps should be performed; such an appreciation would be less immediate with an unfamiliar interface and the user may need specific guidance in how to carry out various steps. However, our evidence suggests that familiarity with the interface is independent of effects of procedural similarity. Consequently, there should be a benefit of using a familiar sequence of steps regardless of whether the interface is the same or not. Of course, a critical ingredient in assessing such design considerations is an understanding of what prior knowledge and experience users are likely to have, and such an understanding is not always easy to acquire. Our view is that this information is essential since designing for learnability means designing for users' prior knowledge.

## Note

1 This research was supported by the Natural Sciences and Engineering Research Council of Canada.

## References

Dixon, P. (1982) "Plans and directions for complex tasks." *Journal of Verbal Learning and Verbal Behavior*, 21, 70–84.

Dixon, P. and Gabrys, G. (1991) "Learning to operate complex devices: effects of conceptual and procedural similarity." *Human Factors*, 33, 103–20.

Dixon, P., Zimmerman, C. and Nearey, S. (1997) "Prior experience and complex procedures." *Memory & Cognition*, 25, 381–94.

Miller G. A., Galanter, E. and Pribram, K. H. (1960) *Plans and the Structure of Behavior* (New York: Holt).

O'Reilly, T. and Dixon, P. (1999a) "The role of functional labels and learning strategy on the transfer of prior device experience." Unpublished manuscript.

O'Reilly, T. and Dixon, P. (1999b) "Procedures are only skin deep: the effects of surface content and surface appearance on the transfer of prior knowledge in complex device operation." *Proceedings of the Cognitive Science Society Conference* (Hillsdale, NJ: Erlbaum).

# 16 The influence of affect on cognitive processes

## Implications of the informative nature of affect in the area of industrial and product design

*Robert C. Sinclair, Sean E. Moore, Carrie A. Lavis and Alexander S. Soldat*

### Introduction

In social psychological research addressing the effects of affective states on judgemental processes, substantial evidence has been garnered that suggests that affective states (see, e.g., Clore *et al.* 1994; Martin and Clore 2001; Schwarz 1990; Sinclair 1988; Sinclair and Mark 1992, 1995) and affective cues (see, e.g., Soldat *et al.* 1997; Soldat and Sinclair, in press) engender differential processing strategies. That is, positive states and cues (e.g. happiness) lead to superficial, less detail-oriented, and more heuristic processing, whereas negative states and cues (e.g., sadness) lead to more detail-oriented/systematic processing. Sinclair (1987) suggested that findings such as these could have particular importance in the area of product design where affective responses to design issues have not been addressed. For the purposes of the present discussion, the terms "design" and "products" are broadly defined. For example, they could include design and product issues ranging from traditional ergonomics to advertising, political campaigns, work space/office design, presentation of people, procedures, to production lines. We believe that affective states (e.g. moods, emotions) and affective cues (i.e. cues that suggest positive or negative valence, but that do not affect the perceiver's mood) caused by, or conveyed by, designs can affect people's performance and perceptions. What sorts of research supports our contentions? We address this issue below.

### Mood states and judgements

#### *Performance appraisal*

As a result of eliciting changes in processing strategy, sad moods appear to enhance analytical thinking and decision-making relative to neutral and happy moods. Indeed, people in sad moods display greater accuracy across many judgemental tasks (e.g. Sinclair 1988; Sinclair and Mark 1995). For example, sad people have been shown to be more accurate in making performance appraisals (Sinclair 1988; Sinclair *et al.* 1998). Specifically, Sinclair (1988) had participants read about a university professor who displayed 16 positive and 16 negative behaviors. The behaviors mapped onto 8 behavioral categories (e.g. preparation and organization, sensitivity, etc.) with 4 behaviors per category. Sinclair varied the number of positive and negative behaviors within category (within subject). Later, he induced happy, neutral, or sad moods in participants and found that sad people were most accurate in their performance appraisals. As illustrated in Figure 16.1, participant's evaluations of the target on scales assessing performance on the 8 behavioral categories mapped more closely onto the behavioral content in the categories (relative to neutral and happy mood participants). Furthermore, sad participants displayed less halo error in their evaluations than did neutral or happy participants. As illustrated in Figure 16.2, sad people discriminated when making their evaluations and displayed lower interdimensional correlations than did happy people (neutral fell in between). Happy people appeared to fail to discriminate across behavioral categories and evaluated the target as positive, neutral, or negative across all behavioral dimensions; this resulted in higher interdimensional correlations. Finally, Sinclair (1988) demonstrated that the global evaluations of the target by sad people were predicted by more behavioral category ratings than were the evaluations of happy or neutral people. Sinclair

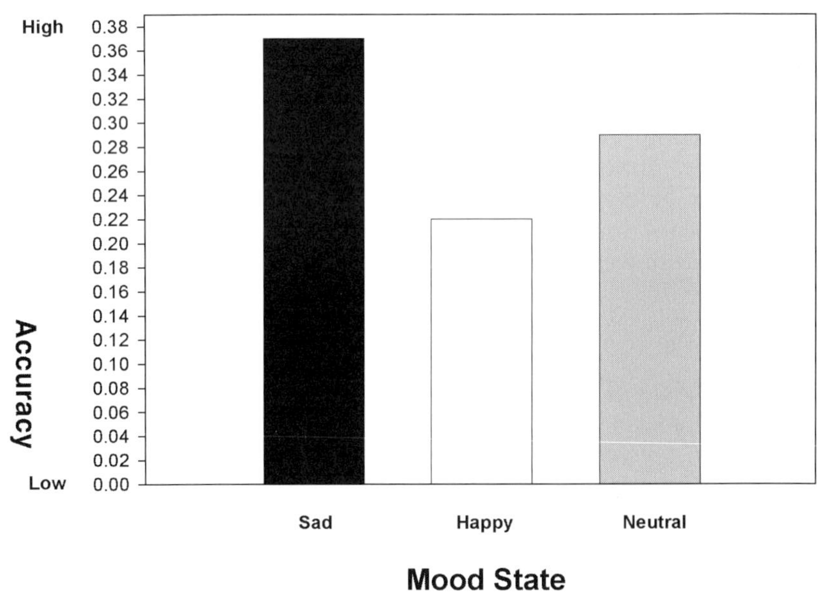


*Figure 16.1* Accuracy of performance appraisals as a function of the appraiser's mood.

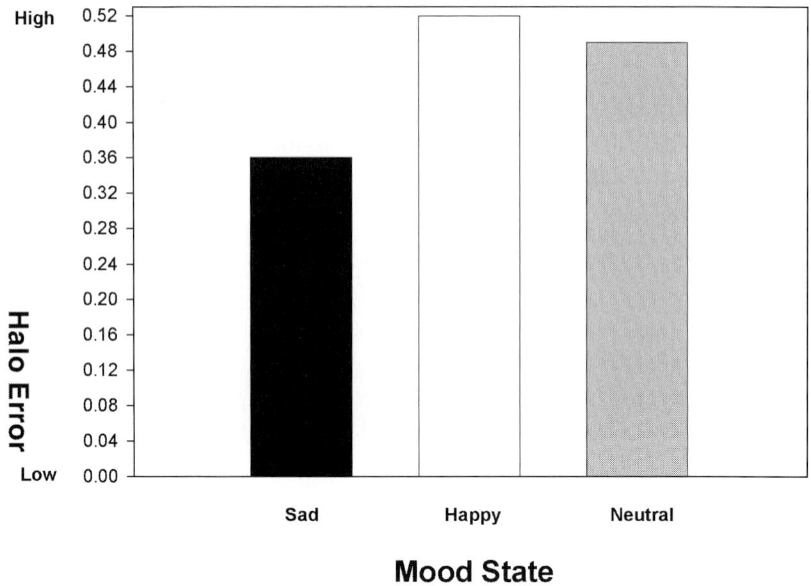


*Figure 16.2* Halo error in performance appraisals as a function of the appraiser's mood.

concluded that sad people were engaging in more deliberate, analytical, systematic processing and were attending to more information and more diverse information when making their judgements (relative to neutral and happy people). Furthermore, Sinclair argued that happy moods caused people to engage in less effortful, less systematic, and more heuristic processing that led to judgemental errors (see also Sinclair and Mark 1992). In the Sinclair (1988) study, the neutral group always fell between the happy and sad groups. Based on the results discussed above, we believe that it might be reasonable to suggest that the design of performance appraisal metrics, and/or the situations in which performance appraisals are conducted, convey negative affect.

### *Statistical Judgements*

Sinclair and Mark (1995) extended the mood-related processing argument to statistical judgements and found that sadness caused greatest accuracy and happiness caused least accuracy. That is, sad people were most accurate in estimating correlation coefficients from scatterplots. Cognitive response analyses indicated that sad people were engaging in more systematic processing than were happy people. As was the case in previous research, neutral control groups fell between the happy and sad groups. Again, there are design issues here, in terms of eliciting the appropriate processing strategies for cognitively complex judgements.

### *Responses to persuasive communications*

In the context of studies addressing responses to persuasive communications, sad people have been shown to be more discriminating than happy people when processing arguments (Bless *et al.* 1990; Sinclair, *et al.* 1994). Thus, sad people elaborate arguments and are persuaded by strong, but not weak arguments. Happy people fail to elaborate and are equally persuaded by strong and weak arguments. Mackie and Worth (1989) and Worth and Mackie (1987) demonstrated similar happy-neutral differences (cf. happy-sad). Thus, happy moods appear to lead to heuristic processing of arguments and reliance on peripheral cues such as source attractiveness, likeability and expertise (Sinclair and Mark 1992; Sinclair, *et al.* 1999). As a result, happy people do not elaborate arguments, and are equally persuaded by strong or weak appeals, whereas sad people elaborate, attend to argument strength, and are thus persuaded only by strong arguments. Thus, from a design perspective in, for example, the area of advertising, it might be important to produce a congruence between the affect conveyed by the ad and the processing goals. If an ad is relying on peripheral cues for product promotion (e.g. likeability, attractiveness, or expertise of the source of the messages; see Petty and Cacioppo 1986), then the ad should convey positive affect; if, however, strong and cogent messages are used in the ad, the ad should convey negative affect.

### *Job performance*

Consistent with our contention that moods may impact various types of judgement, Lavis and Sinclair (2001) explored the effects of mood state on employee productivity. Three studies addressed the effects of affective states on productivity and challenged the common sense belief that "happy workers are better workers." As illustrated in Figure 16.3, in Study 1, sad participants were significantly more productive building circuit boards than were happy participants (i.e. sad people made fewer errors; however, no differences emerged in the number of steps completed on the circuit board task). Study 2 addressed whether happy people were maintaining their moods by failing to devote energy to the task whereas sad people were engaging in affect repair by devoting energy to the task. Happy or sad moods were induced in participants, moods were measured, participants built circuit boards, and moods were measured again. As illustrated in Figure 16.4, Study 1 was replicated and, as shown in Figure 16.5, happy participants' moods did not change from pre to post performance, whereas sad participants' moods changed in the positive direction. Study 3 more directly addressed the influence of motivational factors (i.e. the desire for mood maintenance/affect repair). Happy or sad participants were provided with one of three expectancies about the impact of task performance on their mood state (i.e. mood-maintaining, mood-attenuating, no expectancy).

*Figure 16.3* Errors and steps completed as a function of mood: Study 1.

*Figure 16.4* Errors and steps completed as a function of mood: Study 2.

As illustrated in Figure 16.6, productivity was best for the groups who believed the task would lead to positive moods, and worst for those who believed the task would lead to negative moods, with the no expectancy groups replicating the effects in Studies 1 and 2. Thus, one explanation for mood-related processing effects involves perceived hedonic consequences

*Figure 16.5* Mood change: Study 2.

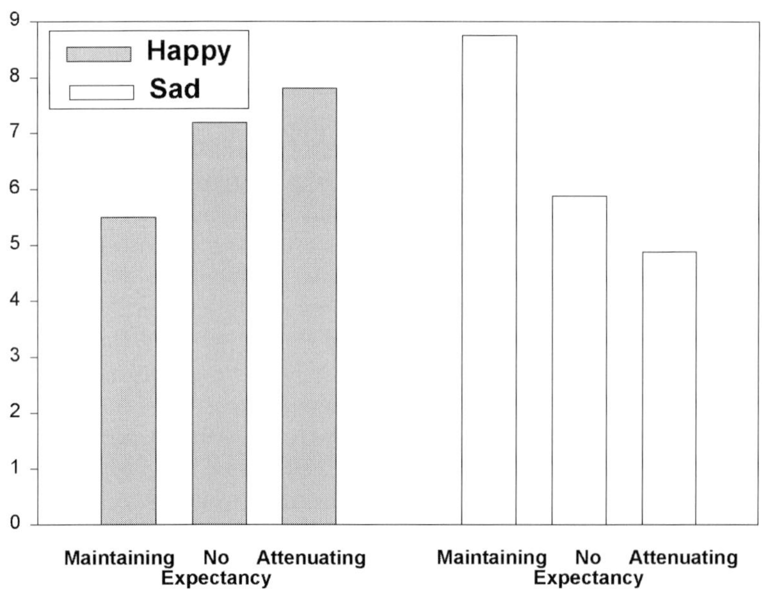


*Figure 16.6* Number of errors as a function of mood induction condition and expectancy: Study 3.

of task performance (see, e.g., Lavis and Sinclair 2001; Sinclair and Mark 1992, for a more detailed explication): that is, people might be motivated by the desire for mood regulation, and attempts to achieve positive mood states lead to differential task performance.[2] In terms of design, we would argue that (a) work situations should be designed to facilitate detail-oriented processing for complex tasks (i.e. convey negative affect), (b) work situations should be designed to facilitate superficial or heuristic processing for simple tasks (i.e. convey positive affect), and (c) work situations that cause people to find happiness in their work might be universally beneficial.

## The informative nature of affect

Another explanation for these mood-related differences in information processing stems from the cognitive tuning extension of the affect-as-information hypothesis (Clore *et al.* 1994; Ottati *et al.* 1997; Schwarz 1990; Sinclair and Mark 1992; Sinclair *et al.* 1994; Soldat and Sinclair, in press; Soldat *et al.* 1997). The affect-as-information hypothesis (Schwarz and Clore 1983, 1988; see also Clore *et al.* 1994; Schwarz 1990), proposes that people use their current mood state as a source of information about the state of their lives in general. Schwarz and Clore (1983) demonstrated that people's moods only affected judgements of life satisfaction when people had no external attribution for their moods. When moods were attributed externally to a nonself-relevant source (e.g. a strange room), the moods were not seen as informative about an individual's life, and there were no mood effects on life satisfaction judgements.

The cognitive tuning extension of the affect-as-information hypothesis expands the informative function of moods, beyond evaluations of life satisfaction, to encompass judgements and decision-making in general. Happy moods lead us to believe that things are going well in our lives, that situations are benign, that we are making good judgements, and as a result, we have no need to pay particular attention to our current situation. Sad moods, however, may act as a signal that things are somehow wrong or threatening, and that perhaps we need to pay more attention in order to improve our current situation. These different types of signals regarding the state of our lives reflect the differential processing strategies used by happy and sad people. If things are going well in our lives, and we are not paying particular attention to things, we are then more likely to engage in processing that reflects an inattention to detail, whereas if we have been alerted to a problem, we will be more vigilant in our processing.

The cognitive tuning framework provides important clues about processes that attenuate mood-related processing differences in judgemental accuracy. If moods are seen as relevant (i.e. originating internally), then they can provide us with a source of useful information. If, however, we attribute the source of our mood to an external factor (e.g. the weather), our mood is rendered uninformative, and typical mood-related processing differences disappear (Clore *et al.* 1994; Sinclair *et al.* 1994).

### *Affective cues*

Recently, Soldat *et al.* (1997) extended the cognitive tuning branch of the affect-as-information hypothesis to external cues (e.g. color) that do not directly influence perceivers' moods. In particular, they argued that colors can serve as external affective cues, thereby signaling the degree of processing required in a particular situation. For example, in Soldat *et al.*, although the colors red or blue had no actual impact on mood, the color red conveyed positive affect and blue conveyed negative affect. Furthermore, mood-like effects on processing were observed in the positive- vs. negative-conveyed-affect conditions (i.e. positive cue led to nonsystematic processing while negative cue led to systematic processing). This pattern of effects emerged in a study in which participants rated their current affective state or the affect conveyed by the paper after solving problems on the colored paper for about 15 minutes. The results demonstrated that participants were most accurate when solving Graduate Record Exam (GRE)-like problems on blue paper relative to red or white. However, paper color did not influence reported affective state (cf. Jacobs and Blandino 1992; Jacobs and Suess 1975) but did influence ratings of the affect conveyed by the paper. An additional study replicated this effect and demonstrated that white fell between red and blue on the measure of conveyed affect (again, no effects emerged on affective states; see Soldat *et al.* 1997). In Soldat *et al.* (1997), another group of participants completed both simple and complex GRE-like questions, involving analytic problem solving, on either red or blue paper and evaluated their current affective states and the difficulty of reading the materials. Blue paper led to greater accuracy, especially for more complex questions. However, paper color did not influence mood or readability. Furthermore, analyses of covariance controlling for arousal (as reported on the arousal component of the measure of current affective state) demonstrated no impact of arousal on performance.

Soldat *et al.* proposed that colors can affect processing strategy, and that the effects of color can be explained by extending the cognitive tuning branch of the affect-as-information hypothesis (Clore *et al.* 1994; Schwarz 1990; Sinclair *et al.* 1994) from experienced affect to external affective signals, such as color. Thus, Soldat *et al.* (1997) argued that positive affective cues signal that a situation is benign and that a processing strategy that conserves cognitive resources (heuristic/nonsystematic processing) is adequate to meet processing goals. Affective cues that are neutral or more negative than usual signal that a person is in a situation where a heuristic/nonsystematic processing strategy is not adequate for dealing with the environment adaptively. As a result, positive affective cues can lead to heuristic/nonsystematic processing whereas negative and neutral affective cues generally lead to more systematic, detail-oriented processing (Sinclair

1988; Sinclair and Mark 1992; cf. Sinclair *et al.* 1999). Consistent with this position, Sinclair *et al.* (1998) demonstrated that students who completed examinations on blue (i.e. affectively negative) paper outperformed students who completed the same examination on red (i.e. affectively positive) paper. This effect was most profound for difficult, relative to easy, questions. Again, the implications in the area of design appear to be rather straightforward. Colors may not affect mood, but can affect processing strategy. Thus, there should be a congruence between the necessary processing goals associated with a product and the design itself.

Other researchers have also found evidence of external affective cues impacting processing without affecting mood. For example, Ottati *et al.* (1997) had participants view a videotape of a person who expressed verbal statements while displaying neutral, happy, or angry facial expressions. These statements could be used to quickly stereotype the person (heuristic processing). Alternatively, these statements could be processed in a piecemeal manner for purposes of judging the person (systematic processing). Among low-motivation participants, neutral facial expressions elicited systematic processing, whereas happy facial expressions elicited heuristic processing. Angry facial expressions elicited a processing style falling between these extremes. Highly motivated participants processed the verbal statements in a systematic manner regardless of the target's facial expression. These results correspond to findings reported in the mood literature (e.g. Bodenhausen *et al.* 1994 who found that positive affective states lead to stereotyping), and the differences obtained for low- and high-motivation participants support a cognitive tuning account of the mediating process (see Soldat *et al.* 1997). Analogously to Soldat *et al.*'s results, no changes in experienced affect were found.

Soldat and Sinclair (in press) extended the Soldat *et al.* (1997) work to other affective cues that did not affect mood and to other areas of judgement. In Study 1, participants were exposed to a set of strong or weak arguments supporting comprehensive examinations for graduating students (Petty and Cacioppo 1984) printed on either red (positively valenced external affective cue) or blue (negatively valenced external affective cue) paper. After participants read the arguments their mood and attitudes toward comprehensive exams were measured. As illustrated in Figure 16.7, the results showed that the blue paper participants elaborated the arguments and were persuaded by strong arguments only, while the red paper participants did not elaborate and were persuaded to the same extent by both strong and weak arguments (note that high scores indicate more favorable attitudes). There were no differences in mood between the groups. In Study 2, under the auspices of a study addressing impromptu speeches, participants read aloud arguments from Study 1 to an audience who responded either favorably (i.e. smile; positively valenced external affective cue) or with a serious facial expression (i.e. somber expression; negatively valenced external affective cue). Later, attitudes were measured. As is apparent from Figure 16.8, participants exposed to the somber cue attended to the arguments and were persuaded by strong arguments only. Participants exposed to the smiling cue did not attend to the arguments and were equally persuaded by strong and weak arguments. In Studies 3 and 4, participants read arguments used in Studies 1 and 2 on a computer screen while photographs of smiling or frowning faces were presented on the screen below threshold of awareness (i.e. a 13 msec presentation). As illustrated in Figures 16.9 and 16.10, the results of Studies 1 and 2 were conceptually replicated. In addition, cognitive response measures collected in Study 3 were consistent with the pattern of results in all of the studies. That is, as illustrated in Figure 16.11, people exposed to the negative cues thought in more detail about the arguments and generated favorable thoughts when the arguments were strong and unfavorable thoughts when the arguments were weak. Finally, as illustrated in Figure 16.12, people exposed to the negative cue recalled more of the information presented in the arguments than did those in the positive cue condition. As discussed in the section devoted to responses to persuasive communications, it appears to be important to design persuasive appeals in a manner that corresponds to the desired processing strategy. Cues should lead people to process in detail when doing so is desirable in terms of the impact of the communication; conversely, cues should lead people to fail to process when relying on peripheral routes to persuasion.

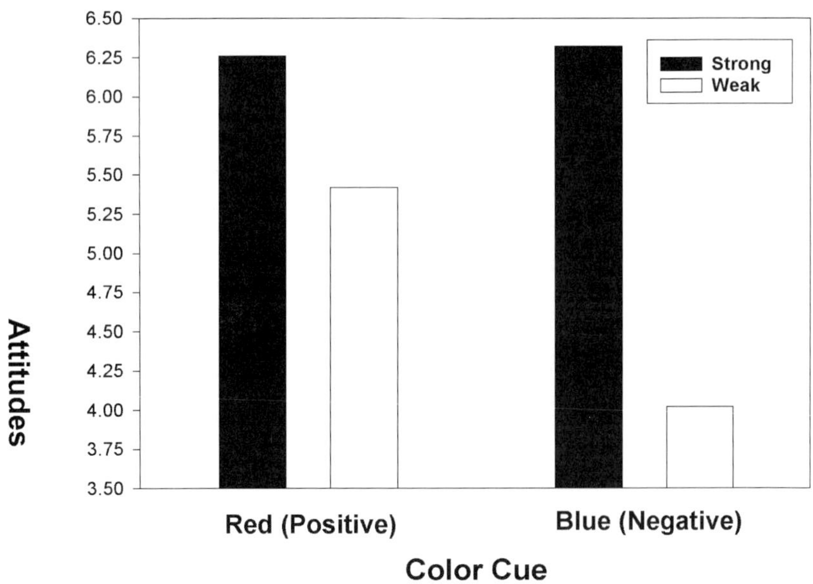


*Figure 16.7* Study 1: Attitudes as a function of color cue and argument strength.

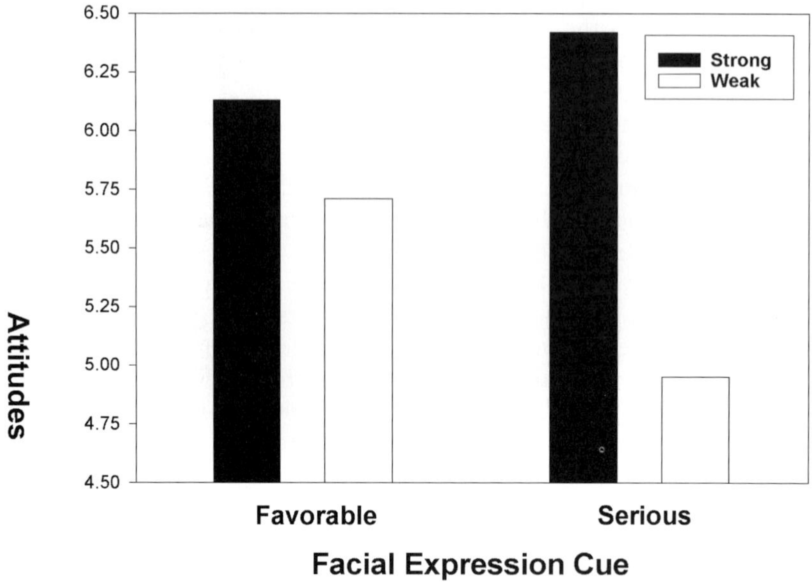


*Figure 16.8* Study 2: Attitudes as a function of audience facial expression cue and argument strength.

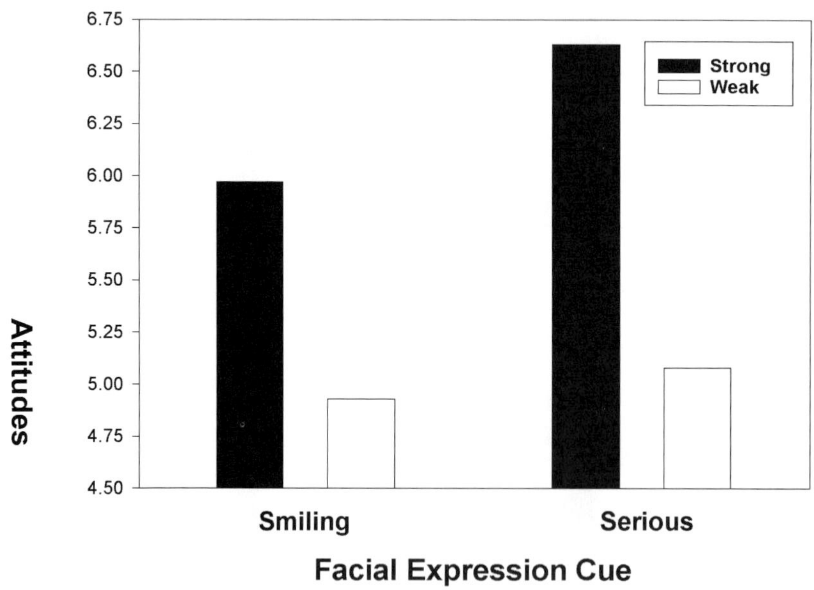


*Figure 16.9* Study 3: Attitudes as a function of subliminally presented facial expression cue and argument strength.

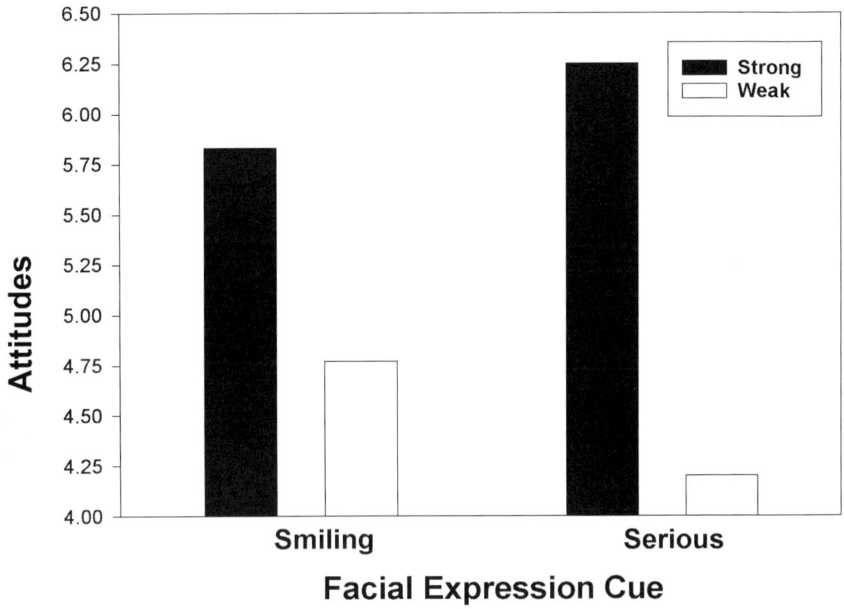


*Figure 16.10* Study 4: Attitudes as a function of subliminally presented facial expression cue and argument strength.

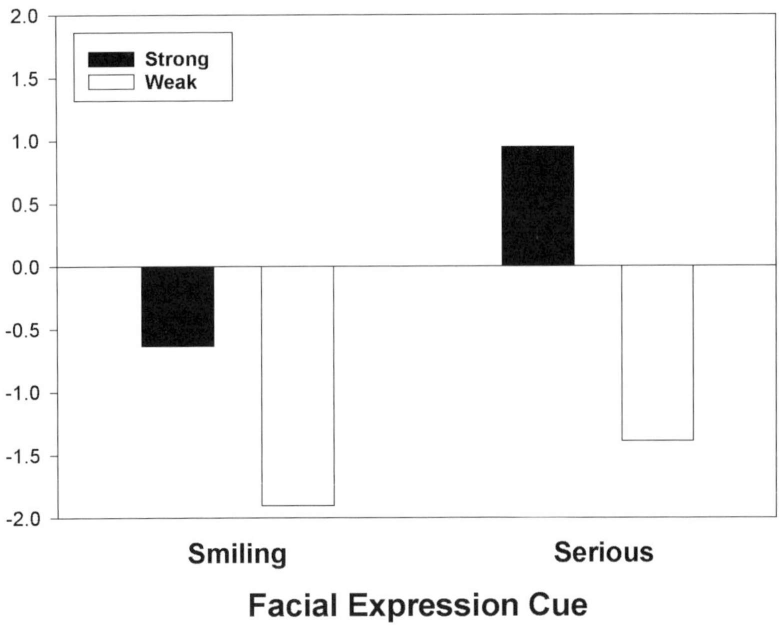


*Figure 16.11* Study 3: Positive thoughts minus negative thoughts as a function of subliminally presented facial expression cue and argument strength.

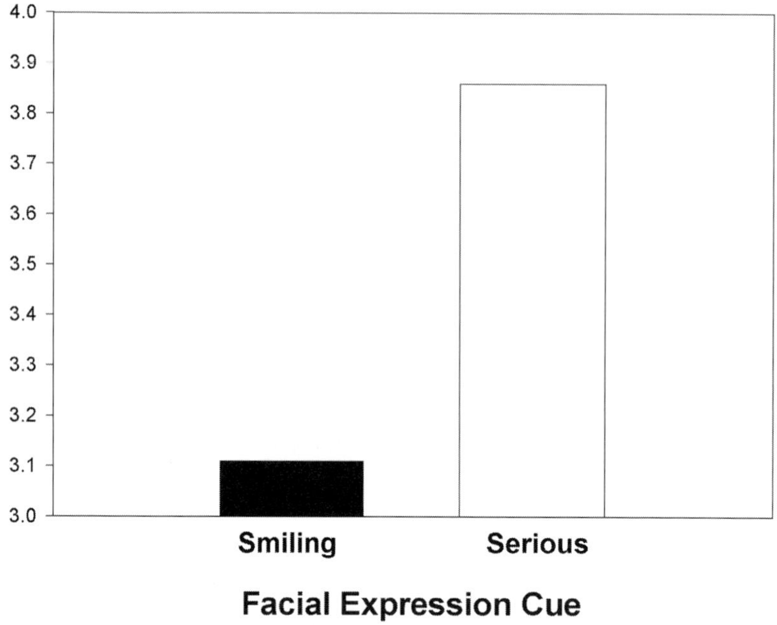


*Figure 16.12* Study 3: Number of arguments recalled as a function of subliminally presented facial expression cue and argument strength.

## Implications in the area of design

Can design lead to differential affective states (or at least lead to differential affective signals)? Obviously, we believe that it can. Indeed, in the area of office design, Larsen *et al.* (1998) assessed the impact of the density of plants in an office on mood, perceptions of performance, and actual productivity task performance. They found that reported mood and perceptions of performance were more positive in the presence of plants, but that actual task performance was inversely related to plant density. This might be indicative of a performance decrement in good moods and a lack of consistency between perceptions of performance and true performance.

Imagine that the design of a product conveys happiness and causes people to process information in less detail. We would expect that this type of design to facilitate processing on tasks that require little effort and that are well learned; however, a design that conveyed happiness might not be the best choice for products or situations associated with tasks that require systematic/detail-oriented processing – here, a design that conveyed sadness might be the more reasonable choice. Given the evidence that affective states and affective cues can affect judgement, decision-making, and performance, we believe that it is time that those involved in design attempt to produce designs that convey the appropriate affect necessary for optimal performance. Finally, though, there is a caveat: certainly an appropriate balance between affective cues, processing requirements, and performance/judgement must be considered. It might be the case that continued exposure to negative affect could lead to product or task avoidance, dissatisfaction, and withdrawal behaviors like lateness, absenteeism, or turnover, whereas positive affect could lead to the reverse pattern (George 1989). Regardless, it seems imperative that those involved in design be cognizant of the affective consequences associated with their designs.

## Notes

1 The writing of this chapter and much of the research reported here was supported by Social Sciences and Humanities Research Council of Canada operating grants to Robert C. Sinclair.

We thank Cor Baerveldt, Jorge Frascara, Doug Olsen, and John Pracejus for their comments on a previous draft of this manuscript.

2 Sinclair and Mark (1992) discuss the causes of mood-related processing strategy differences in more detail. They argue that the effects of mood states on judgment and behavior are multiply determined and involve a) affect regulation (briefly explicated above), b) the informative nature of mood states (explicated below), and c) cognitive capacity. The cognitive capacity explanation of mood effects suggests that happy moods bring more information and more diverse information to mind. This restricts working memory and forces happy people to rely on heuristics. While the capacity explanation does have some support (see, e.g., Mackie and Worth 1989, 1991; Worth and Mackie 1987), more recent evidence suggests that this explanation cannot account for the preponderance of data (see, e.g., Sinclair *et al.* 1994; Soldat *et al.* 1997; Soldat and Sinclair, in press).

## References

Bless, H., Bohner, G., Schwarz, N. and Strack, F. (1990) "Mood and persuasion: a cognitive response analysis." *Personality & Social Psychology Bulletin*, 16, 331–45

Bodenhausen, G. V., Kramer, G. P. and Susser, K. (1994) "Happiness and stereotypic thinking in social judgment." *Journal of Personality and Social Psychology*, 66, 621–32.

Clore, G. L., Schwarz, N. and Conway, M. (1994) "Affective causes and consequences of emotion." In Wyer, R. S. Jr. and Srull, T. K. (eds) *Handbook of Social Cognition* (Hillsdale, NJ: Erlbaum), Vol. 1, 323–417.

George, J. M. (1989) "Mood and absence." *Journal of Applied Psychology*, 74, 317–24.

Jacobs, K. W. and Blandino, S. E. (1992) "Effects of color of paper on which the Profile of Mood States is printed on the psychological states it measures." *Perceptual and Motor Skills*, 75, 267–71.

Jacobs, K. W. and Suess, J. F. (1975) "Effects of four psychological primary colors on anxiety state." *Perceptual and Motor Skills*, 41, 207–10.

Larsen, L., Adams, J., Deal, B., Kweon, B. and Tyler, E. (1998) "Plants in the workplace: the effects of plant density on productivity, attitudes, and perceptions." *Environment & Behavior*, 30, 261–81.

Lavis, C. A. and Sinclair, R. C. (2001) "Mood states and perceived hedonic consequences of task performance affect productivity," Unpublished manuscript, University of Alberta.

Mackie, D. M. and Worth, L. T. (1989) "Processing deficits and the mediation of positive affect in persuasion." *Journal of Personality and Social Psychology*, 57, 27–40.

Martin L. L. and Clore G. L. (2001) *Theories of Mood and Cognition: A User's Guidebook* (Mahwah, NJ: Lawrence Erlbaum Associates).

Ottati, V. C., Terkildsen, N. and Hubbard, C. (1997) "Happy faces elicit heuristic processing in a televised impression formation task: a cognitive tuning account." *Personality & Social Psychology Bulletin*, 23, 1144–56.

Petty, R. E. and Cacioppo, J. T. (1984) "The effects of involvement on responses to argument quantity and quality: central and peripheral routes to persuasion." *Journal of Personality and Social Psychology*, 46, 69–81.

Petty, R. E. and Cacioppo, J. T. (1986) "The elaboration likelihood model of persuasion." In Berkowitz, L. (ed.) *Advances in Experimental Social Psychology* (New York: Academic Press), 19, 124–203.

Schwarz, N. (1990) "Feelings as information: informational and motivational functions of affective states." In Higgins, E. T. and Sorrentino, R. M. (eds) *Handbook of Motivation and Cognition: Foundations of Social Behavior* (New York: Guilford Press), 527–61.

Schwarz, N. and Clore, G. L. (1983) "Mood, misattribution, and judgments of well-being: informative and directive functions of affective states." *Journal of Personality and Social Psychology*, 45, 513–23.

Schwarz, N. and Clore, G. L. (1988) "How do I feel about it? Informative functions of affective states." In Fiedler, K. and Forgas, J. (eds) *Affect, Cognition, and Social Behavior* (Toronto: Hogrefe International), 44–62.

Sinclair, R.C. (1987) "The psychology of product perception." *Interface 87. Human Implications of Product Design* (the 5th Symposium on Human Factors and Industrial Design in Consumer Products, Rochester, NY).

Sinclair, R. C. (1988) "Mood, categorization breadth, and performance appraisal: the effects of order of information acquisition and affective state on halo, accuracy, information retrieval, and evaluations." *Organizational Behavior and human Decision Processes*, 42, 22–46.

Sinclair, R.C. and Mark, M.M. (1992) "The influence of mood state on judgment and action: effects on persuasion, categorization, social justice, person perception, and judgmental accuracy." In Martin, L. L. and Tesser, A., (eds) *The Construction of Social Judgments* (Hillsdale, NJ: Erlbaum), 165–93.

Sinclair, R.C. and Mark, M.M. (1995) "The effects of mood state on judgmental accuracy: processing strategy as a mechanism." *Cognition and Emotion*, 9, 417–38.

Sinclair, R.C., Mark, M.M. and Clore, G.L. (1994) "Mood-related persuasion depends on (mis)attributions." *Social Cognition*, 12, 309–26.

Sinclair, R. C., Mark, M. M., Soldat, A. S., Lavis, C. A. and Moore, S. E. (1999) "Mood state and persuasion: both happy and sad moods can lead to increased elaboration of arguments." Unpublished manuscript, University of Alberta.

Sinclair, R. C., Wuest, S. M., Lavis, C. A. and Soldat, A. S. (1998) "Mood state, true halo, and processing strategy: the effects of mood states on halo and accuracy of performance appraisal judgments." Unpublished manuscript, University of Alberta.

Sinclair, R.C., Soldat, A.S. and Mark, M.M. (1998) "Affective cues and processing strategy: Color coded examination forms influence performance." *Teaching of Psychology*, 25, 130–2.

Soldat, A.S. and Sinclair, R.C. (in press). "Colors, smiles, and frowns: external affective cues can directly affect responses to persuasive communications in a mood-like manner without affecting mood." *Social Cognition.*

Soldat, A.S., Sinclair, R.C. and Mark, M.M. (1997) "Color as an environmental processing cue: external affective cues can directly affect processing strategy without affecting mood." *Social Cognition*, 15, 55–71.

Worth, L. T. and Mackie, D. M. (1987) "Cognitive mediation of positive affect in persuasion." *Social Cognition*, 5, 76–94.

# 17 Gendered spaces of domesticity

*Rosalind A. Sydie*

## Introduction

Space is where power/knowledge relations are realized materially (Foucault 1977). Foucault remarked that, "A whole history remains to be written of spaces—which would at the same time be the history of powers (both these terms in the plural)—from the great strategies of geopolitics to the little tactics of the habitat" (quoted in Soja 1995: 14). The investigation of spaces has, however, tended to concentrate on the great strategies rather than the little tactics. Cities, urban and suburban development, public spaces such as prisons, schools, museums, railway stations, great squares and thoroughfares, have been analyzed in political, socio-economic, and psychological terms, but the "little tactics of the habitat" exemplified by the household or domestic living space have had less attention. It is, however, in those domestic spaces inhabited, on a consistent daily basis, largely by women, that gender power relations are often starkly revealed and played out.

Irigaray would not be surprised at the lack of attention because, as she points out, "In order for difference to be thought and lived, we have to reconsider the whole problematic of space and time" (1984: 15). Her point is that we need to recognize how space and time obliterate the feminine. Irigaray connects this obliteration with the masculine ambivalence towards the maternal womb. She suggests that there is a masculine "nostalgia for the original home, an attempt to keep it for himself, own it, control it, in order to return to it in fantasy (by keeping women in the home, for example, or ensuring their social dependence)" (Whitford 1991: 153). Whilst Irigary's discussion is largely concerned with the space, or rather the lack of space, for women's language and how women can have "their house of language, a home in the symbolic order," her comments also apply to the material spaces of the feminine, most especially to domestic spaces—those "little tactics of the habitat" (Whitford 1991: 156).

Domestic spaces are spaces in which gender relations are materially and discursively organized. Discursive relations are not simply language, they are also material artifacts and sensual experiences—the placing and design of objects, the sensory reception of the spaces and their contents. The meaning of spaces, places and their contents is a language that is embodied in design products. There is a reciprocal relationship between design and meaning when embodied in places and spaces. Bricks, mortar, metal, plastic, paper, textiles and all the other materials of design are all infused with social and cultural meanings. The realization of meanings in design—the shape, form, contour, touch, smell, sight, sound and taste—all contribute to and may transform the particular use of objects. An important component of use is the normative meaning that resides in, and is constructed by, and/or reinforced by, the design of the various spaces and their contents.[1]

Domestic spaces in Western society are infused with normative meanings that reinforce culturally constructed gender configurations. These normative meanings and expectations are reflected in the approach to, and the results of, design practices and products for the home. Because the domestic space is largely a feminine space filled with designs usually produced by men, the normative ambience of that space and its contents is often contradictory for the primary user. In the following discussion, I examine the normative contradictions involved in the notion of "efficient" household appliances and the ideological attributes of the "home."

## Efficient design and domestic space

Household appliances are designed to be both aesthetically pleasing as well as efficient. That is, the artifacts are technically well designed to enable the task for which they are intended to be performed with ease. At the same time, the artifacts, hopefully, have a sensual component, being a pleasure to use, look at, etc. Technical efficiency may not always be matched with beauty and pleasure of use, and both of the attributes may be antithetical to the normative assumptions about the primary user and the space in which the use occurs. Technical efficiency, especially, is often in tension with the idea of domestic space as a moral, protective, emotionally resonant space produced by women.

Domestic spaces are primarily women's domains. Despite the high percentage of dual job or dual career households in Western societies, domestic work and childcare remain identified as primarily a woman's responsibility. Recent statistics on hours spent on housework indicate that, irrespective of employment status, educational achievement, of financial status, women in heterosexual relations still do the bulk of the household labor. Furthermore, the normative assumption is still that this is fundamentally "right" or appropriate. If visitors to a dual job or career household arrive to find the place messy or dirty they do not think, "what a terrible housekeeper he is."

Domestic space is largely a work space for women in contrast to the more leisured space it represents for men.[2] For women, domestic space is the site of everyday/everynight production whereas for most men it is a site of everyday/everynight consumption. Women are not simply involved in the work of housework, they are also expected to produce an emotionally supportive and pleasing space for the other members of the household. The nineteenth-century domestic ideal of the wife and mother as the emotional and moral epicenter of the home remains a potent ideal that underlies the normative assumption that domestic spaces are primarily women's spaces. This assumption is one that, generally, both men and women tend to endorse.

The production of a clean, aesthetically attractive space for household members' enjoyment can be made easier by the use of technically efficient household artifacts. For example, the efficiency of a well-designed vacuum cleaner is a considerable advance over a carpet beater. The problem arises, however, with the normative assumptions surrounding the notion of efficiency when tied to normative assumptions about the ideal home.

"Efficiency" is not a neutral term; when applied to domestic spaces and domestic artifacts it involves certain assumptions about the nature of housework, the capabilities and desires of the houseworker, and the "virtue" of particular domestic practices connected with the use of specific artifacts.

## The nature of housework

Efficiency is about the work of housework. Any activity that is labeled "work" is, ideally, both efficient and satisfying. The problem is that efficiency and satisfaction are not automatically connected. This is certainly the case with much of the work of housework. Whatever the efficient merits, as well as the aesthetically pleasing design, of a machine for cleaning floors, the satisfaction when the task is completed is short-lived—all it takes is someone to walk on that floor.

The idea of efficiency is connected with the efficiencies of industrial production and the assumption that such production techniques (with some modification of scale) are equally applicable to, and desirable for, the home. But the home in Western, industrial societies is supposed to be the place of refuge and relief from industrial, paid labor. The home is quite literally supposed to be the "haven in a heartless world." It is, however, not the space of the home as such that is the haven, rather it is that "women . . . are the haven for men" in the home. The ideology of the home as haven obscures "men's dependence on women and perpetuates the illusion of male autonomy" (Hare-Mustin and Maracek 1994: 61).[3] The key ideological focus is on the relational, emotional work that women do in the home not how efficiently the home is managed.

But women's work in the home is also instrumental, necessarily so if the conditions for satisfactory relational, emotional work are to be accomplished. In this regard it would seem that efficient tools to do the instrumental work would be a benefit. That is, appropriate, well-designed tools would enable efficiency in the work routines that, in turn, would enable greater time and effort to be devoted to the emotional labor of creating the home as "haven."

Making a "haven" is supposedly the creative and individualized aspect of housework. In fact it is this latter aspect of the work of housework that is supposed to make such work pleasurable and desirable. But as early studies of domestic labor illustrated, technical, scientific efficiency does not relate easily to moral and emotional work largely because of the unpredictability of the latter. This is an important source of stress for any houseworker who will have generally bought into the idea that it is her duty, even her "natural" calling, to provide for the emotional needs of the rest of the household members. This duty and/or natural ability—the "creative" aspect of domestic work and thus the most desirable because assumed to be the most "pleasurable"—is also work. And it is work that can lead to frustration because it is inefficient in industrial, productive terms, and the efficiency of the tools that supposedly assist in the performance of these tasks is immaterial to the actual daily practices of emotional and relational support and nurture.

### *The capabilities and desires of the houseworker*

Efficiency is tied, whether in the home or in paid employment, to assumptions about the capabilities of the worker. The assumption is that most workers are rational, cost benefit analysts who are able to grasp the efficiency value of the technique or tool. But cost benefit analysis in paid work is tied to increased production and sales in contrast to the less easily rationalized and estimated costs and benefits of housework such as emotional care and the promotion of optimum individual development for household members. In addition, there is the possibility of concrete rewards for industrial productive efficiency in the form of pay increases or bonuses in contrast to the more nebulous rewards for housework such as love and appreciation. Housework is, after all, supposed to be a "labor of love."

Efficiency, in the industrial production model, also implies that workers are interchangeable. That is, the machines and the techniques for their use are relatively simple so that any worker, irrespective of personal or cultural differences, can use them efficiently.[4] But for relational, emotional work the assumption of interchangeability breaks down. Cultural, racial, ethnic, religious and class factors enter into the issue of the capabilities and desires of the houseworker and how the tasks are performed. These differences also impinge on the acquisition and use of various "efficient" tools to assist in the necessary tasks. For example, a popular sensory signifier of a good homemaker is home baked bread and cookies. It is the smell as much as the taste of the bread and cookies—appearing at opportune times for the other members of the household—that is the key signifier of this prescription. The provision of such an important, emotionally resonant signifier is particularly difficult for the majority of women because of the time constraints that children, paid labor and instrumental housework involve. But there is an efficient tool now available to assist in making home baked bread—the breadmaker.

The breadmaker may be both efficient and desirable as a sign of a good nurturer. It will have an appeal to women who are assumed to be, and generally assume themselves to be, "naturally" more concerned with the care and nurturing of others. However, although the cost may be relatively low, it is still a cost, and a discretionary cost, that may be out of reach for many women. At the same time, to the extent that the houseworker has bought into the idea that home-baked bread, or more generally, home-prepared meals from raw ingredients is a sign of their love for others and their good and virtuous homemaker skills, then the inability to provide these goods can induce stress and guilt.

Instrumental efficiency and emotional, relational tasks are normatively linked to gender in the household and the optimum combination of these tasks is taken as a sign of "virtue," of a woman's unconditional love for others.

## Virtuous practices and products

The houseworker's work is virtuous work when it produces physical and emotional satisfaction for others. In addition, the desire of the houseworker is for emotional reciprocity—the labor of love, ideally, produces love for the laborer—rather than material rewards. As a result, an efficiently designed tool or product is "virtuous" because it enables this critical emotional reciprocity to occur more easily. In reality, the "virtue" of any household product lies in the profits generated by extensive consumption. And such consumption is promoted by the guilt that can be produced in the female consumer. That is, not to use the product reflects a lack of concern and love for other household members. An efficient domestic product may be profitable but, in the last analysis, the use of the technically efficient product is subordinate to the emotional relationships governing the work of housework. The recognition of the potency of the emotional component of housework is what lies behind the success of Martha Stewart and "house beautiful" publications.

## Normative prescription and experience space

Domestic space is home space and it is emotional resonance rather than efficiency that is the most important component of the place designated a "haven." That emotional resonance is bound up with memory and dreams, especially of the first home space we experience and recall. Discussing time–space relations in Bachelard's work, Game uses his point that the "first house, the house we were born in, is physically and psychically inscribed in us" so that "all other houses are but variations on the first theme" to illustrate the power of memory. Furthermore, there is a "passionate liaison" between body and house. "Memories are localized, materially; the materiality of place is inscribed on our bodies" (Game 1995: 202). It is not instrumental work but emotional work that is responsible for these memories, especially for the quality of these memories. And the quality is very much a sensual matter. The smell of furniture polish, the feel of fabric, the sounds of a house, the sight of a "treasure," are the components of memory that underwrite the bodily experienced home. The most evocative memories are often memories of mother that sensory signifiers prompt. These sensory, embodied memories are the significant "little tactics of the habitat" and they are normatively, and in general practice, gendered as female tactics.

## Notes

1 For example, the introduction of a bed in a space labeled "kitchen" would not only be somewhat inconvenient for the usual kitchen activities, it would also call into question the designation of that space as "kitchen." This dissonance would tend to persist even in the face of what might appear to be a "good" explanation for the presence of the bed in this space.

2 It is interesting that the producers of Lazy-Boy chairs have just recently started to target women as well as men as potential consumers, but the design and promotion of the product still suggests male comfort zone—a well-deserved sanctuary after his hard day at work.

3 Hare-Mustin and Maracek also point out that seeing the world of paid labour as a male world also obscures the contribution women make to that world—as support personnel, and as administrators of day-to-day business that enables men's work as managers, administrators and policy-makers to proceed smoothly (1994: 61). Smith (1975) discusses the ways in which women in paid labour often provide the same sort of support and work that wives provide in the home.

4 This is an assumption that often breaks down on the shop floor, for example, when workers resist the dehumanization that accompanies the efficiencies of a production process. Sabotage and slowdowns can represent the assertion of personal control in the face of the alienating demands of the job. The human relations school of management practices believed that when efficiency was directly connected with worker satisfaction then productivity, and thus profits, would increase. Their prescription for linking these two aspects of work was to make the work environment more relational rather than confrontational by encouraging worker participation in the design of products and the manufacturing processes, as well as encouraging a plant or work site community in which workers interacted with fellow workers, including management, on a personal as well as work-related basis. In other words, the workplace was to become a substitute "home" in the interests of profitability and labor peace.

## References

Foucault, M. (1977) *Discipline and Punish*, trans. A. Sheridan (New York: Pantheon Books).

Game, A. (1995) "Time, space, memory, with reference to Bachelard." In Featherstone, M., Lash, S. and Robertson, R. (eds) *Global Modernities* (London: Sage), 192–208.

Graham, L. D. (1999) "Domesticating efficiency: Lillian Gilbreths" scientific management of homemakers, 1924–1930." *Signs*, 1 (3), 633–75.

Hare-Mustin, R. T. and Maracek, J. (1994) "Gender and the meaning of difference: postmodernism and psychology." In Herrmann, A. C. and Stewart, A. J. (eds) *Theorizing Feminism* (Boulder: Westview Press), 49–76.

Irigaray, L. (1984) *Ethique de la difference sexuelle* (Paris: Minuit).

Smith, D. (1975) "An analysis of the ideological structures and how women are excluded." *Canadian Review of Sociology and Anthropology*, 12 (4).

Soja, E. W. (1995) "Heterotopologies: a remembrance of other spaces in the Citadel-LA." In Watson, S. and Gibson, K. (eds) *Postmodern Cities and Spaces* (Oxford: Blackwell), 13–34.

Whitford, M. (1991) *Luce Irigaray: Philosophy in the Feminine* (London: Routledge).

# 18 Social science as a design profession

## New visions and relationships

*Rob Shields*

### Introduction

Amongst other shifts in the social and intellectual context in which social scientists and designers operate, one might list five key shifts in understanding which have important implications for both design practice and social scientific research practice. Although this is not an exhaustive list, highlights of these changes include understanding things as agents, technology as human, nature as artifice, surface as structure, and ideas as objects.

Things as agents: understanding "things" as agents, means seeing them as transformers of peoples' energy and as repositories of latent agency such that a fence may stand-in for a human guard stopping people from entering an area. Objects have the effect of translating both actions and intentions into other actions. Like any agent, they have the power to influence or require changes in others' actions. Breaking down the division between things and people in this manner does not anthropomorphize, but socializes things and places them in a social world as active components. The effect is to challenge separation of the material world as objective and of humans as subjective. This separation has been a political decision with implications for how we have understood our relationship to, and status in, the material world. It also separates the natural sciences from human sciences, constituting academic disciplines in a manner which has been little examined, except for attempts to bridge the separation rather than going back to the roots of the division in the first place (Latour 1999).

#### *Technology as human*

Technology is increasingly integrated with our understanding of human capability, such that, for example, electron microscopes become an accepted part of our collective perceptual apparatus, or artificially synthesized and genetically modified antibodies supplement our bodies' immune system. It follows from examining the complex collectivity of persons, things and animals, that technologies must enter into the equation. Rather than setting technology against a privileged and mystical "human spirit" it becomes possible and permissible to consider their relationship in a more nuanced manner (Woolgar 1988; Law 1991; Callon, *et al.* 1986).

### *Nature as artifice*

The antique division between nature and artifice thus breaks down. Even the most natural environments are understood to be affected—now and in past—by human intervention, whether acid rain or the restructuring of genes. Nature has been one of the untouchable concepts necessary to anchor a division of interests and powers which splits design away from society. However, it is possible to unpack this often unexamined concept to discover that it has a history and a politics (Smith 1984; Eder 1996).

### *Surface as structure*

Post-structuralism and complexity theory have both challenged the Platonic division between appearance and invisible structures which are presented as the "truth" of appearances. When surface is understood as a more direct, and less concealing outcome of structures, then the relationship between objects or events and forces or meanings is revised (Deleuze and Guattari 1983).

### *Ideas as objects*

Finally, considering objects as the embodiment or concretization of values and concepts challenges the division between the ideal and the empirical, problematizing all endeavors built on the separation of action and thought (philosophy), or real objects and ideas (science), or between social laws and everyday life (social sciences). Bringing materialism together with idealism focuses attention on the values expressed by material culture.

## Shared elements and interests

The shifts sketched above suggest a moment of convergence in which design and craft may be being re-socialized; that is, recognizing the shared elements as well as distinctive approaches to the shaping of cultures, materials and institutions. If there are underlying similarities, then the social sciences such as sociology have something to learn from design professions. To begin with, both are concerned with relationships and composition.

However, despite the obviousness of the central role of relationships in any social process, the importance of relationality has often been obscured by an emphasis on discrete, Cartesian, units of analysis. Simply by separating out the elements, the nature of their inter-relationships tends to be lost except to the extent that it is recovered by tracing the changes one element forces on another. The high period of positivism, stretching from the 1920s, has focused on discrete units and elements, with scant attention to theorizing relations between elements in context. However, a sociology of forms and forms of social relation and interaction has existed since the early days of sociology in the mid-1800s. Georg Simmel's sociology of forms identified the types of sociability as the key subject matter of a truly social research endeavor (Kaern *et al.* 1990). This is quite different from the methodological individualism of other traditions of social research which arrives at a social analysis by aggregating the individual behaviors of a group of respondents and projecting the collective form of behavior or preference through the laws of statistical probability.

Simmel's interest focused on the various forms of social interaction and their degrees of intimacy (friendship, neighborliness, civil inattention, small groups), whether authentic (the adventurer), urban (the stranger), or commodified (the prostitute), and the relationship of individuals to the collective (for example, the individual and the new metropolitan cities of the late nineteenth century—Simmel 1950).

The theme of relation, connectivity, and pattern is fundamental not only to social theory but to design theory and practice. Where there is a marked distinction between design and the social sciences has been the critical social science interest in the lack of relations—for example, social stratification and cleavages along class, race, ethnicity, age and gender lines. Studies of marginalization focus on the processes of social exclusion and cultural constructions of certain groups as "other." This social analysis is not merely diagnostic but intimately connected to a history of social engineering and which reconnect the marginalized, oppressed or scorned back into mainstream society, or projects of social transformation which reconfigure social relations on a grand scale (Lazarsfeld 1967; Podgorecki *et al.* 1996).

Admittedly, questions of "composition" have been repressed in social science. In its place, one tends to see an historical stress on developing the means of comprehending and understanding other points of view: cognitive methods for understanding and dialogue (Shields 1996), models of group's taken-for-granted everyday behavior and normative codes (Bourdieu 1971), providing economic assistance or guidance—whether technological (e.g. development assistance), normalizing (e.g. therapies to produce "well-adjusted" people) or pedagogical (e.g. family counseling). Rarely does one find the sort of social design which takes seriously the forms and contexts of interaction such that not only the marginalized but also the dominant groups or "players" are drawn into the scope of analysis. In addition to Simmel, one could give examples such as Marx's critique of unequal labor and exchange relationships in capitalist economies (1998), or Garfinkel's study of the social construction of everydayness (1967).

In recent years, interest in the relational has returned in, to take one example, the anthropology and sociology of consumption. Beyond levels and patterns of consumption, attention has shifted to the ends and motivations behind consumption in advanced societies. Studies of distinction (Bourdieu 1984) and consumption classes or taste cultures have mapped the role of objects and consumer purchases in struggles for social status and in the production of individual and group identities. Groups based on elective affinity, such as friendships based on shared interests, mark these collective interests and shared affinities through symbolic means such as a shared clothing or consumption style. The tribalism of brands echoes and symbolizes a group's recognition of a shared experience. Whether they be music or sport fans, wealthy business people or collectors, common "uniforms" and activities mark their identification with each other as a group. In its most sophisticated forms, one need only hint at these "uniforms"—a single item or purchase hints at the entire collection of objects and values that is entailed in a lifestyle. By making goods widely available outside of the niche markets of those who are traditional or practical users of the goods, consumption allows a blurring of the material culture of peoples' actual life-conditions and their ideal lifestyles. Marx once decried the phantasmagoric quality of commodities in favor of their utility and use-value. In the late twentieth century the symbolic aspects of objects which are a feature of the circulation of images in globalized media are integral to knowledge economies.

Consumption thus plays a role in marking out the formation and relation of interest and status groups, the aficionados of niche markets and "neo-tribes" of all sorts (Maffesoli 1996). Shared experiences are expressed through symbolic objects, styles, actions and places.

## Ethical aesthetics

"Aesthesis" is the old Greek term for such shared experiences as well as the root of "aesthetics." The term goes back to Democritus' notion of perception as a "breathing-in" or consumption of one's surroundings or environment (Nills 1985). The world is a shared, common bond, and the basis of the collective urban association between citizens which gives rise to politics (Bauman 1993).

Groups which are marked by an intrinsic shared experience (not just an external classificatory category) are marked by their group ethos, a relation to each other as a collective and to the world. Ethos separates insiders and outsiders but whether positive or negative, friendly or hostile, both are forms of the "ethical," understood in its philosophical sense. Aristotle points out that ethics denotes the forms of interaction which emerge from situational contexts—they are contingent and depend on the independent judgement of those involved as to how they will interact and relate. By contrast, morals denote universal and transcendent rules of interaction and engagement. Judgement is denied the participants who instead are enjoined to follow rules or laws. We might say, following Maffesoli (1996), Lyotard (1994) and Levinas (1985) that Ethics give rise to an Aesthetics.

This has become an unfamiliar usage of the notion of aesthetics which since the work of Baumgarten in the mid-nineteenth century has been formalized and come to refer specifically to art, enshrined in its own autonomous sphere and separate from questions of justice and of ethics. But even as late as Kant, aesthetics is treated as relational judgement—not a moral question of good versus bad but of beauty and harmony of proportion. In his Third Critique of Judgement (1952), he sought to relate the beautiful to justice and truth.

If we reintroduce the idea of an ethical aesthetics (Maffesoli 1996) we are able to pose questions of harmony and balance between groups, in relation to the environment, and to the Other. This is a crucial and strategic concept at this time as it links consumption to issues of the carrying capacity of ecosystems, resource depletion and economic relations with distant Third World producers. The significance of this to the design professions is that a developed theory of ethical aesthetics allows one to bring a broader range of parameters of "good design" to bear on design and production decisions. Design is crucial because it is concerned with imagining what patterns, uses and forces could exist. If the question of the relation to the Other has dominated contemporary sociological and anthropological debates, "design" has been a missing term in social science.

Ethics alone is insufficient to make changes or guide actions. It is a content that requires a form—an aesthetics. Ethics arises from the aesthesis of shared experience, a relationship to a world, a workplace or to people. Aesthetics alone is equally insufficient, for it leads to an aestheticized politics of manipulation and of form alone without content. But as beauty elicits love (Plotinus 1964)—the aesthetic which draws an ethical response and commitment from us—design is one process and a rare profession that deliberately brings ethics together with aesthetics. However, the importance of this union has been recognized only for the purpose of increasing market share. Can it be used for the collective good? For this, the design professions will have to take seriously the critical insights on the relationship to otherness that the social sciences have developed.

## Social aesthetics: social science as a design profession

Social science interest in the diagnosis of social problems, the administration of policies intended to manage problems, and attempts at implementing solutions has focused on tools such as legal and financial penalties. But there has been less attention to the often more successful efforts of the designers who have sought to attract users and consumers rather than regulate them. For example, direction and information signs in a building may well be backed up by barriers, but the efficient and safe movement of large crowds, such as those that come out of a sports stadium requires that peoples' attention be gained so that they move toward an exit long before they encounter a physical barrier. These efforts to attract peoples' attention and engage them in carefully attending to a message, a set of instructions, or to perform a duty, lies not in the area of regulation and governance but in the tradition of aesthetic experience and of ethical action. The importance of design to the social sciences might be clarified by thinking of this as a social aesthetics—a term which captures the importance of relational, aesthetic experience for the social sciences and the contribution the design professions might make in better understanding peoples' interaction with others—be they other people, animals, things or environments.

## References

Bauman, Z. (1993) *Postmodern Ethics* (Oxford: Blackwell).

Bourdieu, P. (1971) "The Berber house or the world reversed." In Douglas, M. (ed.) *Rules and Meanings* (Harmondsworth: Penguin), 98–110.

Bourdieu, P. (1984) *Distinction: A Social Critique of the Judgement of Taste* (London: Routledge & Kegan Paul).

Callon, M., Law J. and Rip, A. (1986) *Mapping the Dynamics of Science and Technology: Sociology of Science in the Real World* (Basingstoke: Macmillan).

Deleuze, G. and Guattari, F. (1983) *Anti-Oedipus: Capitalism and Schizophrenia* (Minneapolis: University of Minnesota Press).

Eder, K. (1996) *The Social Construction of Nature: A Sociology of Ecological Enlightenment* (Rev. ed.) (London: Sage).

Garfinkel, H. (1967) *Studies in Ethnomethodology* (Englewood Cliffs, NJ: Prentice-Hall).

Kaern, M. Phillips, B. and Cohen, R. (1990) *Georg Simmel and Contemporary Sociology* (Dordrecht: Kluwer).

Kant, I. (1952) *Critique of Judgement*, trans. J. C. Meredith (Oxford: Clarendon).

Latour, B. (1999) *Pandora's Hope: Essays on the Reality of Science Studies* (Cambridge, Mass.: Harvard University Press).

Law, J. (1991) *A Sociology of Monsters: Essays on Power, Technology, and Domination* (London: Routledge).

Lazarsfeld, P. (1967) *The Uses of Sociology* (New York: Basic).

Levinas, E. (1985) *Ethics and Infinity: Conversations with Phillippe Nemo*, trans. R. A. Cohen (Pittsburgh: Duquesne University Press).

Lyotard, J.-F. (1994) *Lessons on the Analytic of the Sublime: Kant's Critique of Judgment* (Stanford: California University Press), sections 23–29.

Maffesoli, M. (1996) *The Time of the Tribes: The Decline of Individualism in Mass Society* (London: Sage).

Marx, K. (1998) *The Communist Manifesto* (Halifax, NS: Fernwood).

Nills, M. (1985) *Morality and Self-interest in Protagoras, Antiphon, and Democritus* (Leiden: Brill).

Plotinus (1964) *Essential Plotinus, The Enneades* (New York: New American Library).

Podgorecki, A., Alexander, J. and Shields, R. (1996) *Social Engineering* (Ottawa: Carleton University Press).

Simmel, G. (1950) *The Sociology of George Simmel*, ed. and trans. K. H. Wolff (Glencoe, Ill.: Free Press).

Shields, R. (1996) "Meeting or mismeeting: the dialogical challenge to verstehen." *British Journal of Sociology*, 47(2) (June), 275–94.

Smith, N. (1984) *Uneven Development: Nature, Capital and the Production of Space* (Oxford: Blackwell).

Woolgar, S. (1988) *Science, the Very Idea* (Chichester: Horwood).

# 19 Two weddings and still no funeral

## Sociology, design and the interprofessional project

*Christena Nippert-Eng*

### Introduction

This chapter is about two interprofessional, student team-based projects that I have conducted bridging industrial and university interests. At the Illinois Institute of Technology (IIT) we call these "interprofessional projects," or "IPROs." Participation in IPROs has just become part of the general education requirements for all of our undergraduates. I was one of the first faculty to conduct IPROs and has served on several years' worth of committees exploring and investigating their possibilities for the future of our institution and our students' careers. To date, I have received four professional awards from IIT that reflect the success of my students, corporate sponsors, colleagues and myself in this arena.

The following discussion begins with some of the background that led me to get involved in this form of learning. Next I report briefly on the structures and outcomes of the two projects. The first was with the Ryerson Steel Coil Pickling Company and the second was conducted with Hewlett-Packard's HP Labs and Home Imaging Divisions. This chapter concludes with a brief discussion of some of my other interdisciplinary endeavors with designers that have emerged from my IPRO experiences, highlighting some of the mutual advantages that occur when designers and sociologists work together.

### Motivational context for engaging in IPROs

#### *Personal work preferences*

First of all, my interests as a sociologist include the sociology of everyday life, culture, work, occupations, knowledge, science, technology, space and time, complex organizations, business and economics, sports and leisure, gender, social psychology, symbolic interaction and cognitive sociology. In other words—and this may be important if one is going to engage in the kinds of projects I'm about to describe – it is hard to find a subject matter or a way of thinking that I don't find inherently interesting.

Secondly, the kinds of work I like to do—the kind of job task mix that I prefer—also is diverse. I like to work alone and with others. I like teaching myself and engaging in internal conversations with other authors through the research and writing process. I also like teaching others and forcing myself to better understand them and my own interests in the process. I enjoy long-term, intellectual endeavors with no immediately apparent, applied use-value and I also enjoy short-term, high risk, more tangible outcomes with immediate, applied and high use-value. This kind of preferred job mix also is important for an academic who wishes to successfully engage in IPROs.

### *The university and broader political economies*

Like any other employee, academics behave the ways they do not only because of their preferences but because of the broader political and material conditions that shape our lives. Shortly after I joined the Illinois Institute of Technology the sociology major was abolished along with dozens of other majors and programs. The new administration merged previously autonomous departments, axed staff and faculty positions and otherwise downsized our institution. Simultaneously, the administration announced a new vision for our school: "interprofessionalism."

This was a strategy that the administration hoped would rescue us from our failing financial status by establishing a brand name in education. Our largely technical or professional school consists of mostly engineers, architects, designers, attorneys, and natural scientists. IIT administrators hoped that interprofessionalism combined with project-based learning would be a way to package our strengths into something unique. I took this goal seriously and decided to do what I could to contribute to our new mission.

Our precarious place in the political economy of the higher education system was not the only concern that led a number of us at IIT to address the idea of project-based learning. The explosion of the computer industry has led to increasing support for project-based, multidisciplinary teamwork, especially in the field. The use of consultants and other contingent labor sources along with "technology transfer" programs and the government-encouraged blurring of academic and industrial circles further contributed to our sense of the possibilities for creating a new teaching venue. If we were going to properly prepare students for a world such as this then we needed to reconsider our options for doing so.

### *Interprofessional, academic-industrial projects: "IPROS"*

Initially, the IIT administration handed out a fairly solid definition of what would constitute an interprofessional project. The criteria that have more or less stuck through the development process are as follows. First, these projects would have both a faculty mentor and an industrial sponsor. Ideally, these people would work together to achieve project goals. Second, each one-semester IPRO would be funded by a contribution from the industrial sponsor. This fee is now $10,000. In practice, however, it frequently has been reduced, waved, or accepted in kind. Third, IPROs would include students from at least two disciplines. This largely is supposed to be where the interprofessional nature of the experience comes from. There are numerous ways to substantially bolster the interprofessional nature of the project, however, as I will demonstrate shortly. Fourth, IPROs would help students learn how to (1) work as a team, (2) produce a solid deliverable at the end of the project, and (3) further develop their verbal and written communication skills. This goal has now been expanded to include further development of electronic communication skills. Fifth, IPROs originally were to address technical (that is, engineering), social science and humanist concerns in roughly equal parts. Paying some attention to matters other than technical ones is still a goal but the social science and humanist requirement is not enforced in any way. Finally, faculty not only are to demonstrate but to teach and evaluate all these things for each student within a three-credit commitment. This is, perhaps, the most extraordinary and problematic part of the IPRO structure.

## Personal variations on the IPRO theme

The two IPROs I have designed and conducted so far begin with three assumptions. First, I believe that ethnographic methods are an excellent way of uncovering problems and systematic patterns of thought and practice. Second, an interprofessional approach to defining and solving problems simply makes the best sense to me. A team approach not only makes the most sense to me, it is certainly the most fun and creative way to tackle a problem.

Why am I so convinced that an interprofessional approach to problem defining and problem-solving is likely to be most appropriate? The answer lies in my third assumption: I think of an object—any object, whether a tangible product, a cultural concept or something as elusive as one's self-identity—as a hyperlink. This perspective means any object—and any behavior or any problem surrounding or incorporating that object—requires a multidisciplinary, even an interdisciplinary approach in order to best understand it, define it, and/or solve it.

For instance, in my book, *Home and Work* (1996), I found that the ways individuals manage their keys are intimately linked to literally dozens of other behaviors and objects that appear throughout daily life. People who keep their home and work keys together on the same key chain are likely to have a highly integrated life. They actively blur a spatial, temporal, behavioral and social psychological boundary between home and work. People who separate their home and work keys onto different key chains are likely to have a strong boundary between these worlds.

As an object, then, I found that one's key chain is linked to numerous other behaviours that we frequently don't even notice like commuting behaviors, appearance management, the way we talk at home and work, office and home decor, and eating and drinking habits. But key chains also are linked to trajectories as diverse as the domestic division of labor, occupational norms of boundary work, the history of industrialization, family composition, and one's position within an organizational hierarchy. Links to more physical factors also are present such as the production and shaping of metals, and norms of who has access to buildings, rooms and cars. It is quite easy, therefore, to think of the keychain as a very inter-professional manifestation or hyperlink.

## Ryerson Steel Coil Pickling project: coil storage and traffic pattern logistics

### *Project history*

The three assumptions mentioned above are reflected in the ways I designed and executed the following projects. The first project took place for the Ryerson family of steel companies. The Ryerson Coil Pickling plant is in the business of receiving, storing, "pickling" and shipping huge coils of steel. The coils are sent to Ryerson by a client via truck or rail and stored in a large parking lot until the client calls to request processing of the coil. When that happens, the coil is retrieved, taken to the plant and unrolled through a hydrochloric acid bath to remove corrosion and allow inspection for the quality of the steel. The coil is then rerolled, wrapped up, placed on a truck or rail car and sent back to the client.

The work we did for Ryerson was one of the first two IPROs to be attempted at IIT. It came about when I met the plant manager, Tom Ziech, at a Discovery Zone birthday party attended by our five-year-olds. He invited me to come visit his plant in order to see his high performance, cross-functional team in action. I quickly accepted and soon visited the plant, located on the south side of Chicago.

After a couple of hours touring and talking, I was settled into the office with Tom and his general manager. I questioned them about a number of things I had noticed, including a possible traffic problem. As I was driving onto the site I got lost because there was no visible signage directing me to the correct building. In addition, once I got directions from personnel in another building, got back in my car and headed in the right direction, my way was blocked by a couple of eighteen-wheel trucks that were delivering and/or receiving the giant 40,000 pound steel coils. My path was blocked again when I tried to leave.

I asked the two men if this happened often and whether or not the back up constituted a problem for the truck drivers as well as the passenger car drivers. The two managers looked at each other, let out a collective sigh of exasperation and said, "Yeah, well, that's a problem we've been having a lot of." Because the plant was operating at a much higher capacity than had originally been expected, the traffic patterning simply wasn't working any more. It was becoming a serious problem as truckers sometimes waited up to six hours to be unloaded, accruing some very real business losses.

The second matter I mentioned was that I still had not been able to figure out how the steel coils were organized in the coil field. This is a very large parking lot, basically, with large hunks of cracked rock over the dirt for a surface. Long lines of coils are arranged in double rows, stacked two coils high each. Each coil had a long, maybe eight digit number spray painted on its side. However, there was no apparent logic in how one coil number related to another even between coils that were directly adjacent. After I asked about this, the two managers gave each other another quick, smiling glance and said, "Yeah, that's another problem." Actually, it turned out that the number on the side of each coil was generated by accounting and had nothing readily visible to do with the size, ownership, or placement of a coil. This made it very difficult for workers to locate a given coil, especially if it had been misplaced. This happened frequently. In order to retrieve a coil with a large "Taylor driver"—a forklift-like machine—the driver often had to move several coils to open spaces elsewhere in the field in order to get at the desired one. If the driver forgot to put one of the shuttled coils back it may not be found again for months. This was especially true since the drivers relied so heavily on their personal cognitive maps of original coil locations in order to later retrieve individual coils. Changes by any of the drivers would not be noticed and committed to memory, which was the only retrieval information that they had. Shortly after my visit to the pickling plant the institutional downsizing changes at my university occurred. IPROs, which no one had heard of until then, became a key focus of administrative efforts. I called up Tom and asked if he would like me to put together a project focusing on the traffic patterns and coil storage problems we had discussed. The plant staff had neither the time to do this themselves nor the extra budget to hire a consultant to solve the problems for them. We both thought it would be fun to work together, too. It was a go.

## Project design

### *Field work*

The one-semester Fall of 1996 project had a number of components. Eventually, I recruited nine exceptional students over seven majors to participate in the work. Most were sophomores; there was one senior.

To begin with, I decided that the students would get a crash course in fieldwork in order to conduct extensive observations, interviews and even some unplanned participant observation. (Hard-hatted and shoed with steel-tipped boots, a number of the mechanical and civil engineers got to ride around in the Taylor driver, allegedly so that they could see how difficult it was to locate and move the coils while in the driver's seat.)

This piece of the project resulted in some very interesting ethical dilemmas, incidentally. These ranged from what we were going to do with potentially damaging information that students had been given during interviews about the actions of some employees to how I was going to handle the interface between the nature of my students and the nature of the work site. On the latter, I committed what I believe is the only overtly sexist act of my teaching career when I refused to allow the two women on the team to interview workers during the nightshift.

### *Contextual research*

The project also included a fair number of lectures by myself and in-house guests in addition to archival work using Ryerson and library and internet documents. For instance, IIT faculty expertise and goodwill resulted in a number of guest appearances. These included a lecture on the economy of the steel industry (which put various pressures on the plant from the way work was organized, to a potential property sale at the site, to its location relative to the rails and major highways, etc.). Early in the process, we also heard about the possible use of technical writing techniques to shape and eventually help present our research findings. One colleague spoke to us about traffic analysis techniques used by city planners. The nature of steel and its constraints on the process were important, too, so we heard from two metallurgists who tutored us on the production of steel and its atomic composition. Ryerson management spoke to us about the ways that self-directed work teams operate. This was important not only to understand how the division of labor at the plant might contribute to and solve the problems, but it raised important issues for our own performance as a team. Finally, pavement options and the territory of the civil engineer were important for our understanding, too, so I set up another lecture to address these issues.

### *Brainstorming for identifying and solving problems*

A third piece of the project design included structuring some serious attention to and time for brainstorming among the team. As we talked—often for an entire week's three-hour class meeting period at a time—we learned how to use collective, multidisciplinary talents to solve a problem. More importantly, we learned how to use them to define the problem itself.

### *Presentation of deliverable*

Finally, the project design included a good bit of attention to techniques of effective communication. This occurred at many levels, including the presentation of our findings. Because of the highly controversial and political nature of IPROs at the time, I invited quite a few faculty and administrators to our presentation along with the top two men from Ryerson. In the end, the students gave a 15-minute, concise presentation to a room filled with nearly forty anxious and powerful people. Students were pelted by questions for over an hour and a half afterwards, demonstrating a remarkable familiarity with the material, the process, and each other as team members. Ryerson, by the way, immediately adopted four of our five recommendations to solve the problems we had studied. The fifth was pending due to the outcome of a potential property sale. We had presented two possible solutions whatever the outcome of the deal. Management said they were highly likely to follow our advice whichever way it went. At least as important an outcome for the Ryerson management, though—I think—was the extent to which these two industrialists enjoyed working with our team.

### *Pedagogical goals of the project*

The details of the substantive problems that we addressed were not the most important part of the project for me. As a teacher, my primary goals for this project were to (1) instill an appreciation of other disciplinary thought-styles in the students; (2) expose students to a positive model of team work and project management; (3) teach students some fundamentals in field work skills, contextual research, and the effective use of brainstorming in collective problem-solving; and (4) help students hone various communication skills by providing a platform in which they could practice them.

I specifically designed the project around these goals rather than the solving of the traffic pattern and coil storage problems per se. It was quite successful on these counts according to extensive course-teacher evaluations, student behavior during the project, and individual and collective conversations with students throughout the duration of the project and afterwards.

## Hewlett-Packard (HP) digital photography for the home: analysis of the PhotoSmart camera, scanner and printer

### *Project history*

I had the same educational goals in my next IPRO endeavor, this time with the added twist of conducting proprietary research. There were no new challenges to the management of students in the field with this project. However, my struggle with the appropriateness of allowing students and myself, as a professor, to do proprietary work extended throughout the project. Eventually I decided that when the interests—and bank accounts—of industrial partners can be used to create truly valuable and unique learning opportunities for students that they would not otherwise have access to, such work is worth the trade-off of not being able to discuss findings with others.

The origin of this work once again demonstrates how one can create interprofessional project opportunities within one's normal, single-investigator, single-discipline research interests. I received a call in January of 1997 from Michel Benard of HP Europe. As technology transfer coordinator for Hewlett-Packard, Michel had read some of my work and wanted to know if I would be interested in doing some work on behalf of HP Labs and the Home Imaging Division. They were willing to buy out my time from my institution to do so. I didn't think that would be agreeable to IIT but was very much interested in the project. So, as part of the 16-month project, I designed a one-semester IPRO for the Fall of 1997 in which students would have a chance to work on the problem, too. We would evaluate the new HP PhotoSmart digital photography suite of equipment—a camera, scanner, and printer designed for home use—providing extensive feedback on how and why users incorporated the equipment into their lives, or not.

### *Project design*

The project included three separate studies: a school study, a family study, and an expert user study. The IPRO students were involved in all three aspects. For the first two, they served as fieldworkers, interviewing and observing all participants. They also constituted the user group in the third study, which was actually a self-study. Here they used the equipment, keeping logs and experiential notes to share with the team. The students' work in both capacities was necessary for a proper understanding of use patterns. It is impossible to understand actual usage of computer technology unless it is encountered in the social and physical environment in which the equipment is intended to be used.

Eventually, I recruited twelve students from six programs of study—seven undergraduates and five graduate students—to form the team. My perspective of an object as hyperlink can be seen in the IPRO via the diversity of the team composition.

### *The HP IPRO Team Members*

Michael Carroll, Senior, Computer Science; Jim Esenwein, Senior, Mechanical Engineering; David Fockel, Senior, Design; Frank Gruger, Senior,

Photography; Tayo Ihimoyan, Sophomore, Electrical Engineering; Melanie Joh, Graduate Student, Design; Bill Kerr, Graduate Student, Design; Jay Melican, Graduate Student, Design; Melody Roberts, Graduate Student, Design; Rebecca Trump, Graduate Student, Design; Amy Wiese, Senior, Psychology.

Again, I expected that there would be a number of relevant concepts, trends and cultural behaviors that we would need to know about in order to understand what we would see. Some of these factors included: the effects of gender and age on computer technology purchase and use; the nature of the domestic division of labor especially regarding gift-giving and kin work; the culture of printing; the history and expectations of personal photography; even the nature of the HP organization; team work; the overall computer industry; and the kinds of collaborations that are forming.

The reading list for the IPRO further reflects these links. We all read the first four of these books. Students took turns reporting on the remainder to the class (listed in the order that we discussed them) so that we would have some knowledge of these works.

### *The HP IPRO reading list*

R. Emerson, R. Fretz and L. Shaw, *Writing Ethnographic Fieldnotes*, (University of Chicago Press 1995); H. Wolcott, *The Art of Fieldwork*, (Alta Mira Press 1995); E. Tufte, Envisioning Information, (Graphics Press 1990); D. Packard *et al.*, *The HP Way: How Bill Hewlett and I Built Our Company* (Harper Business 1996); R. Cringly, *Accidental Empires: How the Boys of Silicon Valley Make Their Millions, Battle Foreign Competition, and Still Can't Get a Date* (Harper Business 1996); N. Negroponte and M. Asher, *Being Digital* (Vintage 1996); W. H. Davidow and M. Malone, *The Virtual Corporation: Structuring and Revitalizing the Corporation for the 21st Century* (Harper Business 1993); D. Burstein and D. Kline, *Road Warriors: Dreams and Nightmares Along the Information Highway* (Plume 1996); T. Kidder, *Soul of a New Machine* (Avon 1995); B. Latour and C. Porter, *Aramis or the Love of Technology*, (Harvard University Press 1996); S. Turkle, *Life on the Screen: Identity in the Age of the Internet* (Touchstone 1997); S. Papert, *The Children's Machine: Rethinking School in the Age of the Computer* (Basic Books 1994); R. Cowen Schwartz, *More Work for Mother: The Ironies of Household Technology* (Basic Books 1983); C. Couch, D. Maines and S.-L. Chen, *Information Technologies and Social Orders* (Walter De Gruyter 1996); J. Rothschild, *MacHina Ex Dea: Feminist Perspectives on Technology* (Elsevier Science 1983).

Class meetings for this IPRO—again, held once a week for three hours at a time—included several guest lectures, too. These talks focused on personal and digital photography, the ethics of teamwork, and the technology of the equipment. In addition, we took two field trips. The first was to the Museum of Contemporary Photography in Chicago for a talk on digital photographs in the artistic world. The second was to the Museum of Science and Industry in Chicago, where the director gave us a tour of the "Imaging" exhibit he helped design occupying nearly the entire floor of the huge west wing of the building. The beginning of our semester constituted a crash course in fieldwork. After three weeks we continued to discuss the techniques and ethics of fieldwork but in a different, somewhat more engaging format. This focused on extensive, collective analysis of the videotaped activities of the middle school students participating in the school study.

## School study

Part of this study funded an "after school enrichment program" for one academic quarter each at the University of Chicago Laboratory Schools' Middle School (sixth through eighth grade) and High School. The Middle School program was especially important for the IPRO students' experiences. Here, they began by conducting extensive, one-hour interviews with each of the eighteen students enrolled in the "Digital Darkroom" program. Thereafter, a minimum of two IIT students were present each day of the week while the younger students pursued projects of personal interest using the four suites of equipment and personal computers provided for them. We ended this program with a joint pizza party, celebrating how much we had enjoyed learning from each other and the surprisingly close friendships that developed between both groups of students.

## Family study

The family study extended well past the one-semester duration of the interprofessional project but the IPRO students were critical in launching this part of the research. Thirteen families were selected to receive complete sets of equipment in exchange for sharing their reactions to it with us over a nine-month period. A team of two IPRO members delivered the equipment to each family. They immediately conducted a one-and-a-half hour interview with the person designated by the family as the one most likely to be the primary user of the equipment. They held a 30-minute interview with each of the other members of the household.

Afterwards, we would take copious notes and photographs from the moment the family installer began opening the boxes to the end of the installation process and production of the first photograph. This process included opening the CPU and installing a SCSI card, by the way, quite a challenge for people who had never done any servicing of their computers, nor seen anyone else do so. Students also conducted follow-up telephone interviews within the next month to see how things were going and to remind families to fill in their log books when they used the equipment.

By the time of this project, it had become a formal, campus-wide requirement that all IPRO teams present their findings to judges and all interested parties in a one-day event held at IIT. Of course, our presentation focused on the research process we followed rather than our findings, per se. Of the sixteen projects that semester, my team walked away with the awards for the Best Interprofessional Project and Best Formal Presentation.

## Outcomes of the Ryerson and HP IPROs for various stakeholders

### *Students*

The students involved in these interprofessional projects gained a number of advantages. First, the chance to learn through problem-based or project-based platforms is quite exciting for most students. Issues and lessons appear to be more real, more immediate, and more memorable when they are encountered in this forum. Second, the chance to learn and engage in high quality fieldwork was unusual for most students and enjoyable for all. Third, this kind of endeavor created much closer relationships among team members than classroom learning seems to, fostering bonds between students and myself, as well as with the outside groups whom we were observing. Mentoring and friendship rather than lecturing and anonymity was the norm. Fourth, students (and I) gained quite a remarkable appreciation of what individuals in other disciplines can do for you. The ability to understand and take advantage of other disciplinary thought styles is a lesson not to be taken lightly in the current world of work. Fifth, students in both projects received a great deal of recognition and awards for their work, ranging from breakfasts with the Board of Trustees, trips to conferences on interprofessional work, and cash prizes. Finally, though, and perhaps most importantly for them at this time in their lives, these students landed some remarkably good jobs. The majority of them have reported back to me that they believe their experience in the IPROs helped them stand out far from the rest of the crowd of applicants. It also provided them with additional skills that let them hit the ground running once they began their new positions.

### *Institutions*

Uniformly, Ryerson Coil Pickling, Hewlett-Packard and IIT have been delighted with the results of the IPROs. Both firms approached me immediately to do another project for them at the conclusion of the ones reported here. Like Ryerson, Hewlett-Packard also began to immediately put the research findings into effect, none of which would have had the depth and breadth of understanding for which I was credited without the work of the IPRO team.

### *Self*

I gained a number of skills, insights and other rewards from these projects. First, I began to acquire and hone my abilities in project management, personal time management, team management, riding the industrial–academic fence, and interweaving scholarly insights within work geared towards a deliverable with a deadline. By the second project, I got better at these things than I ever would have imagined. Second, I was able to address and settle a number of pedagogical concerns and interests. How can you create a meaningful experience for students across a variety of interests and levels of education and produce a deliverable and keep it to a 3-credit commitment? Is this really a good teaching venue? What is it best used for and how? How can I simulate its advantages within other kinds of courses? I had a number of long-standing and new ethical concerns, too, ranging from managing students and the sensitive information they were bound to learn in the field to the proper link between industries and universities. Was this really using students as cheap, contingent labor? Should their work be "owned" or simply used exclusively by a private firm looking to make a larger profit based on it?

Finally, in terms of my institutional outcomes, the most important one for me is the sense of satisfaction I have gained in having met the challenge of doing this kind of work—and doing it better than a number of others who have tried it. This reward is closely tied for first place with the sense of satisfaction I feel from the close relationships I have had with my students and the career opportunities that they say that they have received as a result of their IPRO experiences.

From the outside world, particularly the design community, Ryerson, and of course, HP, my rewards have been enormous. Ryerson's Tom Ziech and I continue to be friends and our families get together socially fairly often to talk about everything from kids to work. For the work I did in addition to and surrounding the IPRO, HP paid me very well. They intend to pay me better in the future, too. The next project will be for a very different division on a line of products that does not yet exist. I have acquired a marvelous network of colleagues and a model of what the work process should be from HP, too. It is filled with good, smart people who make wonderful colleagues and collaborators, especially Michel Benard, my original contact, and Elaine Parchman, the head engineer on our project. Michel and I plan to co-author some work on our collaboration, too.

## When sociology and design come together

Perhaps the biggest reward for me, though, has been my much stronger link with the world of designers. First of all, I now see design as the essential, crucial link to any interprofessional project. I have several other projects in mind and they will begin with a strong design presence complemented by other disciplinary members. But I've concluded also that within the other layers of my professional agenda, finding additional opportunities to be useful to and learn from designers is at the top of the list.

### *What sociologists bring to the table*

When I first heard about user-centered design, or design ethnography, I wondered just what sociologists might offer to designers. I have decided that we bring to the table at least the following useful things: (1) a distinctive conceptual, analytical framework; (2) contextual information via substantive areas of interest (in my case, things like everyday life, time, space, culture, work, home, social psychology), including a way of looking at the relationship between people, objects, and activities—especially the politics of design; (3) ethnographic skills; (4) teaching skills in the classroom and in the field, based on some rather rigorous training found in most sociology graduate programs; (5) a different kind of group leadership, perhaps, than what designers are used to; and (6) writing skills. In short, sociologists are experts on the external influences that encourage people to behave the ways that they do. If that isn't useful to a designer, I don't know what is.

### *What designers bring to the table*

But I also was very curious to find out what designers might offer to sociologists. Any time a relationship is not based on a mutually beneficial foundation, it is doomed to failure. Fortunately, it quickly became apparent that I had lots to gain from hanging out with designers.

First, colleagues and graduate students are always nice to have. These make the world go round. And for me, I've acquired these relationships not just within the Institute of Design at IIT, of course, but within every organization that our alumni go to work for, like IDEO San Francisco, E-Lab, Steelcase and Doblin Group. These firms are full of fascinating people and fascinating projects.

Second, design offers a stimulating intellectual potpourri, where answers and work count. For scholars who need this sort of thing, the chance to talk about and assist in a choice collection of real world problems that need real solutions, real fast, is just priceless. It is nothing less than astonishing for someone used to endless, fruitless committee work to see how fast the business world actually values and implements one's assessments and suggestions. Third, designers offer me the chance to better flesh out my own understandings of areas in which I am supposed to be an expert—re: theory, practices, and methods—and new ways of pushing the envelope in these areas. For example, designers are much more tuned in than are sociologists to the physical relationships we have with the objects around us. I've always thought of myself as being very good at the symbolic, emotional, and habitual side of this: what the object stands for, what it evokes in us, how it fits in to our routines and practices. But I've realized that I cannot possibly understand that without also understanding the corporeal, physical, ergonomic, or sensual interactions that we have with an object. This is the designer's domain—at the very least. Methods courses I've developed specifically for design students help push me as well. I think of design ethnographers as the equivalent of Ethnographic SWAT (Special Weapons and Tactics) teams. These courses not only make me put my ethnographic money where my mouth is, but they help meet my need to work with smart, hard-working graduate students and to do this in a compelling, high-level context.

Another wonderful opportunity I'm pursuing with designers is the "Center for Ethnography and Documentary Research" (CeDoc), Kelly Costello's brainchild located in Chicago. This is an organization founded to promote and explore ethnographic methods and their use across industry, government and academia. The Board of Directors consists of sociologists, anthropologists, psychologists, a folklorist, user-centered designers, documentary photographers, film-makers, and a human rights attorney. We're from top universities and liberal arts colleges, corporate firms and non-profits such as Doblin Group, Intel, SonicRim and the ACLU.

I also gain additional awareness of current problems within my areas of interest that I should address in my own work by working with designers. Talking with graduate students, attending conferences like this one and the IDSA meetings this summer are all great ways to find out the hottest problems and concerns in the world of work. I was invited to speak at the IDSA courtesy of Rebecca Trump—an HP IPRO student—and her colleague, Aura Oslapas at IDEO San Francisco (the awards ceremony is probably the closest I'll ever get to attending something like the Oscars).

Opportunities like this are so valuable for raising my awareness of trends, problems and successes that aren't discussed in other arenas and other professions. The book I'm completing now on telecommuting, for instance, is filled with insights and issues that are a direct result of my contact with the design community.

Employment, especially consulting, is an obvious and important opportunity that the design community offers to sociologists, too. Consulting helps me to meet not only my preferences for a large task mix and for meeting all kinds of new people doing all kinds of interesting work, but also helps in the area of material conditions. I'm doing my best to see that other sociologists realize the possibilities for full-time employment here, too.

If I can offer better writing skills to designers, there is no question that they give me some really pretty and effective graphics. I now have an acutely painful awareness of how much I've neglected my skills in and understanding of visual communication. My entire discipline could benefit from incorporating and addressing the basic visual narrative techniques of designers into our work.

Finally, designers offer me—and for now, any other sociologist—a kind of "Special" status. This is something that I think every academic and author, maybe every sociologist seeks. At the moment, I'm one of a handful of sociologists I know who have caught on to the possibilities of working with designers. For now, we constitute a very small club tackling a wonderfully diverse collection of problems together and appreciating the very different but very complementary gifts that we bring to the table every time we sit down.

## Conclusion

Based on my interdisciplinary experiences I've drawn a few conclusions about what it takes in order for these things to happen and to happen right. These are the attributes that I think one must have in order to achieve truly remarkable outcomes. These also are the attributes of the people that I select to work with—whether students, academic colleagues, or industrial representatives.

I look for people who honor, or at least respect, the unique offerings that anyone brings to the table, whether that person is a student, an industrial partner, or a colleague from a different discipline.

I look for people who enjoy the process of discovering what our unique offerings might be and figuring out how we can complement each other in order to get the job done.

I look for people who can stay focused on the work. This is why it makes so much sense to tackle something like this in a project-, problem- or case-based effort. This forum lets everyone access and keep their attention on the problem rather than things that exert too much centrifugal force on a team. And I like people who like to work hard. I make sure that no one on the team works harder than I do and that the demands are reasonable. Slackers are not welcome, though. They exert too much of a destructive influence on the collective.

Finally, I look for people who enjoy celebrating the unique, synergistic —and correct—things that we can accomplish when we choose to work with and learn from each other in this way.

## References

Nippert-Eng, C. (1996) *Home and Work* (Chicago: University of Chicago Press).

# 20 Design and the social sciences

## *The working groups*

In addition to the papers presented, four working groups met to discuss different aspects of the relations between design and the social sciences. Following are the summaries prepared by the working group leaders.

### Working group 1: Design and the social sciences in the university (working group leaders: Desmond Rochfort and Rosalind Sydie [report editor])

Initial questions: How does the university facilitate or does not facilitate a convergence between the social sciences and design? What do we expect to be the outcome in the 4th year of the Bachelor of Design program with the Social Sciences pathway?

Some ideas: (1) theory and practice must go hand-in-hand in the design of the curriculum; (2) we must relinquish the usual teaching/lecture format and provide the student with the ability to ask "why" and "what are the consequences." (This emerged out of the point that often with design education, the initial question: "Why design?" and the final question: "What is the impact/consequences of the design?" are simply not part of the design equation. Similarly, for the social scientist in the academy, the why and the possible outcomes are theorized but the actual practice—the definitive product—is ignored.); (3) the student should, at the end, have a reflective, critical focus on material culture.

Material culture: critical reflection on material culture. The social consequences of actions, whether they are design or social practice, provides the guiding focus of any formal proposals for academic training. Critical reflection involves an ethical stance towards any practice. For example, for design the injunction of environmental responsibility should be a central component. Similarly, for the social sciences the injunction of "do no harm" should be a central component.

There was considerable criticism of the buffet approach—bits of design and social science but no depth to the exposure and little attempt to make connections between the various social science disciplines and between the various approaches to design. There was a general agreement that there are ways to be found to integrate knowledge at all stages of the program.

Final but by no means conclusive end point: the first degree program will produce a reflective, responsible generalist, able to understand the critical issues animating design and the social sciences. This student is not a specialist; specialization is the function of further degrees.

## Working group 2: Design and the social sciences in industry (working group leaders and report editors: Liz Sanders and Jim Wilson)

A small group of conference attendees (about 15) took part in the Design and Social Sciences in Industry Working Group. Nearly all the participants were involved both at the university and in consulting or industry (or they aspired to do so). About half were designers and the other half were psychologists, ergonomists or human factors practitioners. It is interesting to note that no anthropologists or sociologists joined the group. We met twice.

The workshop participants jointly determined the issues we discussed. The moderators took notes and facilitated the discussion such that the highest priority issues were adequately covered. Seven general issues were raised over the two sessions. Five issues were discussed and are described on the following pages. Two issues were noted as worthy of later discussion.

### *Design for individuals*

A student participant began by addressing a question to the industry practitioners in the group: "How do you design for the individual in Industrial Design and what about specific cultural groups?" A discussion about the changing nature of relevant segments emerged, i.e., from demographic market segments, to lifestyle segments, to niches, etc. The issue of social equity followed, i.e. "How can a culture produce products for their own countries when they have to compete with large multinational companies?" Several participants from large multinational companies stated their company's position on this issue.

### *Competitiveness and sharing*

The contrast between the university and industry in terms of competitiveness and sharing of knowledge was brought up. The general consensus was that sharing worked well in academia but not very well in industry. Perhaps because most of the participants had a role both in academia and industry, the group agreed that more sharing in industry was desirable, pointing out that sharing could create the conditions for a company's future survival. We talked about what was considered sharable in industry, including: (1) frameworks and models; (2) processes; (3) methods; (4) tools (sometimes); (5) technologies (sometimes). We all saw the need to keep new technology and specific product innovation a proprietary matter. Some argued that processes, methods and tools should be more readily shared since it is really the people (who use the processes, methods and tools) who should be considered one's competitive advantage. Someone mentioned that industry and academia should work together more so that industry's attitude toward sharing might be influenced by academia's attitude toward sharing.

### *The relationship between industry and academia*

We decided to brainstorm the ways that industry and academia could connect and/or collaborate with each other. The following ideas emerged.

Industry could provide to the university: (1) contracts; and (2) grants. The notion of academic freedom was brought up in relation to industry's providing grants and/or contracts to the university. Some people felt that industry-sponsored and industry-determined projects would rob professors of the right to pursue their own research interests. Others felt that professors had a choice as to whether or not to work on grants and/or contracts. A related issue, that of industry's dictating the university's curricula in return for the hiring of graduates, was also raised. Members of the working group felt that this was not a positive situation in the relationship between industry and academia. Industry could also provide: (1) paid internships (where students perform meaningful work); (2) paid coops (again, where students perform meaningful work). Practitioners could come to the university to share what they know with students and faculty through formats such as adjunct professorships, sessionals, lecturer positions, occasional reviews and/or critiques, as well as invited talks. Lastly, industry could provide: (1) scholarships for students; (2) competitions where students could win either money or be offered paid internships; (3) site visits; (4) participation in conferences and collaborative workshops; (5) chairs; (6) professorships; (7) industry associate programs; (8) loans; (9) equipment and/or tools.

It was pointed out that many university students are not currently aware that there are applied jobs available for applied social scientists.

The university could provide to industry: (1) access to expertise; (2) consulting expertise in the form of consulting groups made up of faculty, undergraduates and graduate students. These consulting groups could be either focused on a specific field of knowledge/expertise or could provide a relevant skill-based team; (3) professional development and lifelong learning opportunities; (4) the loan of professors on sabbatical; (5) students for "live projects" (i.e. industry-sponsored projects); (6) site visits; (7) participation in conferences and collaborative workshops; (8) students who publish work performed while on coops, interns or live projects; (9) the opportunity to "mine" the professional literature.

### *What industry is looking for*

The working group concurred that the most important "deliverable" the university should offer to industry was the graduate. While the industry practitioners did not feel it was necessary for graduates to know the specifics of the design development process, they did agree on the following abilities: (1) writing; (2) communicating; (3) presenting and persuading; (4) teamwork; (5) people skills; (6) EQ (emotional quotient); (7) storytelling ability; (8) self-knowledge; (9) self-direction; (10) time management skills; (11) problem solving ability; (12) process skills; (13) creativity; (14) innovation.

### *Ethics*

Concern was discussed that some products and services may harm individuals and/or environments, and that industry may not be adequately or appropriately addressing these. If so, what role can, and should, universities play in this? How can you ensure that the industry that you do work with (and maybe also industries you don't work with) behaves appropriately? Many industries already address some of these ethics concerns through internal training.

How is "good ethical design" promoted by industry? Is it just used as a form of "marketing," or will industry develop better ways to make ethical actions more profitable?

Finally, in her speech, Christena Nippert-Eng raised a related issue: When academics work with industry, there is often a need for industry to keep the work and its output secure—if so, what will academia do?

### *A common language*

A point was raised that we should be setting our goals toward the "collaboration/integration" end of the scale instead of toward the "connection/cooperation" end of the scale. The working group did not have enough time to address this issue, but felt that it was important to address it at some later point.

### *Connections between all the disciplines*

The absence of anthropologists and sociologists in this working group led to the observation that making connections within the social sciences and within the fields of design was just as important as making connections between design and the social sciences. It was noted that the relationships might be different in academia and in industry. The working group did not have time to address these discussion points, but felt that they were important to address at a later point.

A final point/question was raised: "Where is marketing at this conference?'

## Working group 3: Design and the social sciences in research (working group leaders and report editors: Sharon Poggenpohl and Ausra Burns)

### *Similarities and differences*

Beginning the discussion by identifying similarities and differences between the social sciences and design, participants sought to understand how research and practice in these areas correlate. Historically, design and the social sciences come from different traditions. Design started as an applied discipline primarily focused on the production of objects, structures and artifacts, while the roots of the social sciences lie in theory and research that are applied as social policy. As a discipline, design is not as well codified, institutionalized or as broadly recognized as the social sciences. Design seeks actionable knowledge while the social sciences seek defensible knowledge. On the other hand, both design and the social sciences are accountable, evolutionary and innovative; both are engaged with audiences or users; and both deal with complexity through context relations.

### *General definitions*

Some broad working definitions for the social sciences and design anchored the discussion. The social sciences are a repository of knowledge about human abilities to think, behave, communicate, create culture and evolve values. Design is action and problem-solving oriented, and is prescriptive. It turns abstract ideas into tangible artifacts that serve social/cultural purposes. Design synthesizes artifacts and conditions derived from research to alter/reinforce human action.

### *Research strategies and goals*

Several key components which shape social science and design research were identified during working group discussions. It was the consensus that, in their research, designers and social scientists observe, describe, analyze, understand, reflect, synthesize, predict, recommend and provide advocacy. Overall goals of research recognized by the social sciences and design include hypothesis testing and validation, description of observed phenomena, generation of theory and applied knowledge, creation of more informed and reflective action paradigms and adaptation to new challenges. Social science and design goals are based on social values as well as values which are specific to each field. Research methodologies in design and the social sciences have different aims and functions depending on research applications. In design, methodology is often justified by validity of a result, while in social science the method validates the result. Design methodology is of two kinds: research, and product generation that is specific to design as a discipline. The relation between these two methodological approaches is highly interdependent within design research.

Certain commonalities in the nature of research in the social sciences and design enable both designers and social scientists to build more integrated collaborative practices. Knowledge generated within design and the social sciences does not forge clearly predictable results or precalculated certainties. In both areas this operational knowledge is generated with a similar goal: that of going beyond avoiding failure in research toward application of findings in the social arena. Design and social science research is context limited, but does deal with complexity in terms of multiple, interacting factors that should be considered in focused research. Several directions were identified toward strengthening this synergetic relationship between design and the social sciences. Designers in their professional practice can offer suggestions for studies to be made by social scientists. They can make social scientists' ideas tangible through prototypes. In turn, social scientists can work to reveal human behavior and social contexts that impact on behavior. Social scientists and designers have an opportunity to work in a complementary relationship while generating and applying social agendas. Overall, collaboration can lead to understanding of how products and communication influence behavior.

### *Research collaboration*

To make such research collaborations productive and mutually satisfying, practitioners and academics in fields of social science and design should seriously reconsider their agendas toward establishing broader notions of what design and social research and practice encompass. Design and the social sciences need to negotiate a common vocabulary to ensure effective communication. Representatives from various areas of expertise should seek to expand their contacts with willing collaborators from other disciplines. Designers and social scientists might then create models that open possibilities for effective research and practical design work that also support a common goal of building knowledge.

Designers in particular need to work collaboratively on addressing particular issues arising in societies rather than limiting themselves to the production of objects. They must achieve a cultural shift that recognizes design research as a necessary component of everyday design. Design research can become integral to the designer's intellectual growth and self-enrichment, rather than only be seen as a tool to increase profits and efficiency of products. Designers can build a research culture by seeking public exposure, making their research accessible and desirable, publishing findings under a design research rubric, and educating the public about the various forms of knowledge generation in design.

To encourage effective collaborative research in design practice designers should seek connections with other researchers who focus on problem-oriented projects. Building a body of knowledge is perhaps a more pressing issue than the research itself. Currently most design research falls between disciplinary cracks. A codified body of knowledge would strengthen interdisciplinary research funding. Designers and social scientists can create opportunities for work on fundable, focused projects within academia and industry by seeking institutional support. A supportive work atmosphere and such allowances as course releases can contribute significantly to the advancement of such research and project work.

Reconsideration of the education process within design and interdisciplinary graduate studies in general is of paramount importance toward creating a base for collaborative research. Designers need to develop, integrate and communicate clearly research criteria at all levels of the education process. Along with expanding the notion of the role of the designer, a revised criterion for candidates for design programs needs to be developed and related to a set of research expectations. At the same time, design programs should become more open to students coming from other fields.

## Working group 4: Interdisciplinary cooperation (working group leaders: Jeanette Blomberg, Christena Nippert-Eng [report editor] and Rick Robinson)

The working group on interdisciplinary cooperation focused much of its discussion on the identity and autonomy of participants in interdisciplinary endeavors. These concerns manifested in issues ranging from: (1) the motivation and outcomes for participants who seek interdisciplinary, cooperative endeavors; and (2) the competing and sometimes antithetical ways of working, thinking and rewarding people associated with a given profession or employer, to (3) how to best prepare students for interdisciplinary work.

The group began its conversations by adopting the proposition that integrative, interdisciplinary design is the only approach to design that works. Therefore, the question we decided to address was "how do we achieve this?" Rather than seek a single answer to this question we would try to identify key issues that need to be considered while attempting to create, support and maintain interdisciplinary cooperation. The problem of creating an environment where interdisciplinary cooperation can flourish has no single solution. Rather, effective solutions will have to be crafted (most likely in an iterative manner) for each particular situation.

Examples from our own experiences were helpful not only in identifying important issues to consider, but also in suggesting possible ways to move forward. The following collection of insights emerged from discussion of those experiences. We developed a preliminary list of things that we need to do if we are to foster interdisciplinary work.

### *Communicate the value of interdisciplinary cooperation*

First, we need to find ways of articulating and communicating the value of interdisciplinary cooperation. Interdisciplinary cooperation can: (1) result in more comprehensive solutions to problems because of the multiple perspectives that are represented; (2) change understandings of what the problems are; and (3) provide a greater range of both practical and analytical tools to our deliberations. As we move more and more into designing services and not just stand alone products, it becomes increasingly important to bring in a wider range of perspectives to bear on problems. Certainly, one way to communicate the value of interdisciplinary cooperation is through examples of projects that have worked, showing how those collaborations have paid off and were rewarding for the participants.

### *Communicate who we are to other project participants*

In order to build mutual understanding and trust, participants in interdisciplinary projects need to develop ways of communicating who they are and what they can offer to the problem-defining and problem-solving process. There is wide variation within a given discipline (e.g. design, anthropology, sociology, art history, etc.) concerning such things as backgrounds, skills, experiences, and values. Participants shouldn't assume others know what they can or would like to contribute to the collaboration.

### *Maintain who we are among ourselves*

Discipline-based identity and authority are important. Few people are eager to give up their identities as representatives of a particular discipline. Few institutions would allow them to do so even if they wanted to. While some individuals are not so tied to titles and prefer to simply use whatever works in order to solve a problem, most people appear to feel that their value at the interdisciplinary table stems from their backgrounds in particular disciplines. Maintaining this identity does not in any way diminish others' identities unless we privilege one discipline over another. This is an issue of power, not the functional nature of disciplinary identity, per se, and one that warrants careful policing in interdisciplinary endeavors.

### *Recognize similarities across disciplines*

It is important to recognize similarities across disciplines. At some level the social science and design disciplines share common concerns. For example, we are all concerned with the relations and interactions among people and their material worlds. We share aesthetic sensibilities and ethical concerns. Moreover, when it comes to the interventionist impulse of design, applied social science also is committed to intervening, making things different/ better. In these cases the problematic is often formulated as the design of something new (e.g. processes, delivery systems, political strategies, etc.). It seems that collaborations with designers would be quite useful in these applied situations.

### *Acknowledge and adjust conflicting reward structures*

Inside the university there may be competing standards and reward structures across academic units and throughout different levels of the organization. Tenure reviews, standards about "real" publications and teaching, the need to maintain certain levels of full-time-enrolled students put different pressures on faculty. In other words, accountability to departments, deans, and granting agencies put additional competing interests into the picture. Because of this, outside money may be the best way to support research by interdisciplinary teams, circumventing institutional and disciplinary stakeholders. Also, and ironically, all of these factors may mean that being interdisciplinary in the academy can be a very lonely and unrewarding proposition. A great deal of attention must be paid to the kind of reward structure that will support faculty in becoming more interdisciplinary, teaching interdisciplinary courses, team teaching, and cross-department teaching, etc.

Within the interdisciplinary team, there is a further need to recognize everyone's contribution, while at the same time get away from an emphasis on individual achievement. The interdisciplinary team needs to be recognized for the work they could only have done together. For students, faculty and employees, institutional reward structures that focus on individual effort must be adjusted.

### *Foster project-based learning*

Project-based learning requires attention to a number of factors. First, consistent with the last point, the value of working together should be recognized by those working together as well as by outside funders/ supporters. Project work requires mutual respect and trust. It requires that we learn enough about other perspectives so that we can collaborate effectively (e.g. language, tools, values, dreams, etc.). It requires techniques for constructively bringing the group back to the problem at hand or renegotiating the problem over time. And learning how to be really good at project work requires being taught by example. Faculty who cooperate in teaching projects are essential if students are to learn good project work practices and ways of thinking. Second, there should be no privileged or permanently dominant discipline in interdisciplinary project work. However, the balance of power (relative weight and commitment) afforded to particular disciplines may vary from project to project or over the course of a project. Especially within a given problem different disciplines probably will take the lead at different times.

### *Consider whether different strategies for interdisciplinary work are needed inside and outside the academy*

Some of the factors that may differ across the industrial–academic divide and that may affect the nature of interdisciplinary endeavors include: (1) who is funding the work (e.g. grants, clients); (2) what kind of outcomes are expected (e.g. solutions, explorations, theoretical insights); and (3) the possible duration and timing of the work.

### *Consider the opportunities for interdisciplinary efforts afforded by client work*

Client work often presents opportunities to define the problem space. The definition of the problem is part of the work. That is, problems are not given strictly by the client but are negotiated with the people who are working on the project area. In the negotiations it is possible to open up more interesting spaces for cross-disciplinary collaborations. For example, a problem could be re-specified as not simply a design problem, but one requiring knowledge about the experiences of the people who might use such a product or service. Or a problem could be re-specified as more than a "know your customer" problem, allowing the researchers to look instead at the possible impacts of new configurations of products and services on the experiences of customers.

### *Preparing students for interdisciplinary work*

Does interdisciplinary training work? Does it water down the disciplines? Is this question different or answered differently between the training of students and their careers in practice? Where is the "inter" located in "interdisciplinary education?" In the individual student? In the teams in which students work? In each of the faculty members who teach them? Across multiple faculty located in multiple departments? There may be a real need for interdisciplinary faculty and courses rather than making the students "inter-professorial." (For instance, maybe taking a little psychology, anthropology, sociology, design with different faculty members doesn't make for such a good interprofessional education.)

Are the problems that students solve interdisciplinary? Is it the nature of the training/curriculum? How? There are a number of approaches to the curriculum for designers, for instance, including case-based curricula, pre- and specialist curricula, liberal arts-type degrees culminating in a design agenda, or curricula with much more design experience within a less traditional liberal arts curriculum. Interdisciplinary team/project work is highly desirable, though, especially since as practitioners in industry problem complexity and time to market motivate use of interdisciplinary teams. The answer to all of these questions may well rest in how a particular group answers the question of what design is, as a profession. Professions are not necessarily disciplines. Perhaps design is a set of values and, as a profession, can pull from multiple disciplines without being one. In part this may be because design keeps both the approach to problem-solving and the model of a good result open which is not the model in academic disciplines. As one participant put it, perhaps "design is an adaptive behavior in search of validity."

To be a designer, it is not clear whether one needs to be interdisciplinary or have a discipline or simply have an identity. However, one does need team skills, communication skills, mediation/conflict resolution skills, and an orientation toward sharing rather than arguing. Designers also bring to the table wonderful imaginations regarding what could be; any training must foster this in design students. Clearly, in these and in other areas, social science members can learn from their design colleagues. Cross-study in all directions enriches all of our disciplines and careers. And while there is a solitary, textual emphasis in social science analysis, there is a group, visual, physical emphasis in the design approach. The combination of the two ways of thinking and working is the most effective.

In fact, perhaps generating a list of desired learning outcomes that may be achieved in any way is the most fruitful way to figure out what interdisciplinary training should look like. More comprehensive, "correct" solutions are likely to emerge from this, adapted to each educational environment within the constraints and resources that are available. While our particular outcomes may vary there is no question that the members of this working group will continue to pursue ever more fruitful collaboration between design and the social sciences.

# 21 Conclusions

## Design and the social sciences, a reconnaissance

*Jorge Frascara*

Design is concerned with the conception and production of objects and systems that contribute to our daily lives. The social sciences' aim is the study of that very life, sometimes aiming at just understanding and discussing it, sometimes at assisting those involved with affecting it. This book attempts to map an initial territory of intercourse in which some concerns are addressed and some difficulties are outlined. The way in which the relation between design and the social sciences will develop depends on the people who implement that relation and on their cultural values. Businesses are interested in profit, and they would use all the knowledge available from the social sciences to optimize that profit. Design is concerned with the conception and production of objects and systems that facilitate certain tasks, permit the realization of others that would be impossible without design and technology, and contribute to the enjoyment of life through the use of those objects. The issues to contend with are several, and they center on cultural choices and physical limitations.

The usual growth expectation of businesses is self-centered and disregards the carrying capacity of any system. The great challenge for this century that has just begun is how to harmonize a limitless ambition for wealth on the part of businesses with the physical limits of the planet and the psychological limits of the exploited countries that sustain the wealth of the super-industrialized minority. The connection between design and the social sciences should be seen two ways: on the one hand we have the possibility of using the social sciences to maximize the business success of the companies that develop design to their highest levels. On the other, we have to keep in mind that, in addition to short term goals and profit expectations, design and the social sciences must look at broad contexts and long-term goals, and develop every individual effort keeping at the top of their agenda the highest and at the same time most basic working ground: the welfare of humanity.

The connection between design, anthropology, psychology and sociology is a good beginning. Conscious as we are about the need for an interdisciplinary approach to the confrontation of any design project we must now look into the administrative organization of knowledge in our institutions of higher learning and reconfigure the departmentalized/compartmentalized structures that separate us from the realities we have to contend with.

We hope this publication will contribute to the transformation of existing epistemological and administrative structures by offering a discussion of the possibilities that more integrated ways of using knowledge can offer to the confrontation of today's challenges.

# Index